IN PRAISE OF DOUBT

IN PRAISE
OF DOUBT

HOW TO HAVE CONVICTIONS
WITHOUT BECOMING
A FANATIC

PETER L. BERGER

AND

ANTON C. ZIJDERVELD

HarperOne
An Imprint of HarperCollinsPublishers

HarperOne

HarperCollins books may be purchased for educational, business, or sales promotional use. For information please write: Special Markets Department, HarperCollins Publishers, 10 East 53rd Street, New York, NY 10022.

HarperCollins Web site: http://www.harpercollins.com

HarperCollins®, ✦®, and HarperOne™ are trademarks of HarperCollins Publishers

FIRST EDITION

Library of Congress Cataloging-in-Publication Data

Berger, Peter L.
   In praise of doubt : how to have convictions without becoming a fanatic / Peter Berger & Anton Zijderveld. — 1st ed.
      p. cm.
   ISBN 978–0–06–177816–2
   1. Belief and doubt. 2. Certainty. I. Zijderveld, Anton C. II. Title.
   BD215.B47 2009
   121'.63—dc22

                                                            2008055098

09 10 11 12 13 RRD(H) 10 9 8 7 6 5 4 3 2 1

CONTENTS

ACKNOWLEDGMENTS

The idea of this book came out of a project of the Institute of Culture, Religion, and World Affairs at Boston University, which had been directed by Peter Berger, now a senior research fellow. The project was entitled "Between Relativism and Fundamentalism." An international working group, consisting of American and European scholars of religion, sought to delineate such a "middle position" from the viewpoints of different Christian and Jewish traditions. The papers from this project will be published separately. While the project dealt only with the religious aspects of the relativism/fundamentalism dichotomy, participants soon realized that there were very important moral and political implications. Specifically, while the participants of the project generally agreed that religious faith could accommodate doubt—that is, one could have faith in the absence of certainty—they recognized that people could and did make *moral* judgments with a high degree of certainty, judgments that frequently had political consequences. But how could

viii ACKNOWLEDGMENTS

religious uncertainty coexist with moral certainty? This question went beyond the agenda of the aforementioned project, which is why Berger decided to write a book that would address both the religious and the moral/political aspects of a "middle position." He asked Anton Zijderveld, who wasn't a member of the working group, to join him as coauthor, because he wanted to work on this topic with someone who had greater philosophical expertise. (Zijderveld holds doctoral degrees in both sociology and philosophy.) The collaboration, with each chapter jointly authored, has been both productive and pleasurable.

The authors wish to express heartfelt gratitude to David Kiersznowski, who generously funded the original project and also made it possible for the two authors to meet and work together on the book, once in Amsterdam and once in Boston.

Wenn wir die Zweifel nicht haetten,
Wo waere dann frohe Gewissheit?

If we did not have the doubts,
Where then would be joyful certainty?

—GOETHE

IN PRAISE OF DOUBT

1

THE MANY GODS
OF MODERNITY

Just before the dawn of the twentieth century, in tones of passionate conviction, Nietzsche proclaimed the death of God. Today, a little over a hundred years later, this prophecy hardly seems plausible. Whether God does or does not exist in cosmic reality is another question. And this question cannot be answered by the empirical sciences: God cannot be the object of an experiment. But in the empirically accessible reality of human life today, there is a veritable plenitude of gods competing for the attention and allegiance of people. Nietzsche thought that he stood at the beginning of an age of atheism. Right now it seems that the twenty-first century is marked instead by polytheism. It looks as if the many gods of antiquity have returned with a vengeance.

The more radical thinkers of the Enlightenment, particularly in France, anticipated the demise of religion in a spirit of gleeful anticipation. Religion was perceived as a grand illusion, one that had given birth not only to a multitude of superstitions but to the most monstrous atrocities.

The wars of religion that followed the Protestant schism in Europe certainly gave credence to this view. Thus Voltaire's cry, "Destroy the infamy!" applied not only to the Catholic Church—in his experience, the mother of all atrocities—but to religion in general. Protestants continued to execute heretics and burn witches with all the enthusiasm of their Catholic adversaries. Nor could one find more appealing religious traditions outside divided Christendom.

The instrument that was to destroy religion was, of course, reason. In reason's cool light, the illusions of religion would evaporate. This expectation was dramatically symbolized when the French revolutionaries enthroned the goddess of reason in the Church of the Madeleine in Paris. This Enlightenment faith did not end with the French Revolution. Indeed, in different versions it has continued to this day. In the nineteenth century that faith was particularly invested in science. Reason, it was thought, would find an inerrant methodology to understand the world and, ultimately, to construct a morally superior social order. In other words, Enlightenment philosophy had morphed into empiricist science. The prophet of *that* mutation was Auguste Comte, whose ideology of positivism had an immense influence on the progressive intelligentsia of Europe and beyond (notably in Latin America, where the Brazilian flag is still emblazoned with the Comtean slogan "order and progress"). It was Comte, not so incidentally, who invented the new science of sociology.

As that science developed, it bore less and less resemblance to what Comte had had in mind. It increasingly saw itself not as a system of philosophy, but as a science based on empirical evidence and subject to empirical falsification. Three thinkers are commonly seen as the founders of modern sociology—Karl Marx, Emile Durkheim, and Max Weber. There were great differences among these three. But when it came to religion, each one, albeit for different reasons, believed that modernity was bringing about a steady decline. Marx and Durkheim, both children of the Enlightenment, welcomed this alleged development. Weber, on the other hand, contemplated it with melancholy resignation.

In the sociology of religion, as it developed in the twentieth century, this association of modernity with a decline of religion came to be known as "secularization theory." This theory proposed that modernity, both because of the spread of scientific knowledge and because modern institutions undermined the social bases of religious faith, necessarily led to secularization (understood as the progressive decline of religion in society and in the minds of individuals). This view was not based on some philosophical rejection of religion, but on various empirical data that seemed to support the view. (Significantly, many of these data came from Europe.) It should be emphasized that this theory was "value-free" (to use a Weberian term). That is, it could be held both by those who welcomed it and by those who deplored it. Thus there were any number of twentieth-century Christian theologians who were far from happy about this alleged process of secularization, but who took it as scientifically established fact with which both churches and individual believers had to come to terms. A few theologians found ways of actually embracing it (such as the proponents of the briefly fashionable "death of God theology" in the 1960s—a wonderful case of "man bites dog").

WHAT IS THE CURRENT STATE OF SECULARIZATION IN THE WORLD?

It's fair to say that secularization theory has been massively falsified by the events of the decades since World War II (which, of course, is why most sociologists of religion, with a very few holdouts, have changed their mind about the theory). As one looks over the contemporary world, it's not secularization that one sees, but an enormous explosion of passionate religious movements. For obvious reasons, most attention has been given to the resurgence of Islam. But the militant advocates of holy war, who are causing the attention, are only a small (though very worrisome) component of a much larger phenomenon. Throughout the vast Muslim world—from North Africa to Southeast Asia, as well as in the Muslim diaspora in the West—millions of people are looking to Islam to give meaning and direction to their lives. And most of this phenomenon has little to do with politics.

Arguably an even more spectacular development is the global expansion of Evangelical Protestantism, especially in its Pentecostal version. In 1906 a revival took place in Los Angeles—the so-called Azusa Street Revival—led by a charismatic black preacher whose fiery sermons rapidly built an interracial congregation. Soon members of that congregation began to "speak in tongues" (the defining marker of Pentecostalism). As missionaries from Azusa spread out across the United States and abroad, Pentecostalism gave birth to a number of growing American denominations. But the most dramatic explosion of global Pentecostalism occurred after World War II—in Latin America, in Africa, and in various parts of Asia. Today it's estimated that there are about 400 million Pentecostals worldwide. This is surely the most rapid growth of any religious movement in history. In addition to the growth of Pentecostal churches proper, there's also what has been called "Pentecostalization"—that is, the growth of charismatic "speaking in tongues," healing, and other "gifts of the spirit" in various Protestant and even Catholic churches. Nor is Pentecostalism the only form of Evangelical Protestantism that has been spreading globally. It's been estimated that there are about 100,000 Evangelical missionaries active worldwide—many from the United States, but others from Latin America, Africa, South Korea, and elsewhere in the world. There's also the broader category of "popular Protestantism"—that is, groups that aren't commonly perceived as Protestant, but whose religious and social characteristics have a Protestant flavor. The most successful of these are the Mormons, who have also grown rapidly in many developing societies around the world.

The Catholic Church (arguably the oldest global institution), hard-pressed in its home base in Europe, continues to be robustly healthy in other parts of the world. The same is true of some of the churches derived from the Reformation—notably the Anglican communion, which has been severely marginalized in England but is doing very well in Africa. Eastern Christian Orthodoxy, after years of persecution by Communist regimes, is also experiencing a genuine revival, notably in Russia.

Indeed, the same can be said of every other major religious tradition. Orthodox Judaism has been growing in the United States and in Israel. There has been a revival of Hinduism, challenging the secular definition of the Indian state. There are strong Buddhist revival movements, some of them engaging in missionary work in Western countries. Japan has seen a number of powerful religious movements, some of them offering syntheses of Buddhism, Christianity, and Shinto. Confucianism, both as a religious and as an ethical system, has been rediscovered in China and in the Chinese diaspora.

There are two exceptions to this picture of a furiously religious world. One is geographical—western and central Europe, the one important part of the world in which secularization theory appears to be plausible. The other is sociological—a thin but very influential class of intellectuals who indeed represent a global secularism. The reasons for these exceptions cannot be explored here. However, to avoid facile generalizations, we need to stress that both cases are quite complicated. While there has been a great decline of church-related religiosity in western Europe (among both Catholics and Protestants), all sorts of religious activity can be observed outside the churches, from various forms of New Age spirituality to charismatic movements. Furthermore, the renewed presence of Islam, which had a strong European presence for hundreds of years in the early Middle Ages, has led to a renewed debate about the Judeo-Christian roots of the much-vaunted "European values." As to the secular intelligentsia, there have been vigorous religious revivals in this very stratum, especially in the non-Western world. Thus it's often the children of highly secularized intellectuals who suddenly come out as adherents of this or that militant religious movement.

In sum: It cannot be plausibly maintained that modernity necessarily leads to a decline of religion. Some late descendants of the radical Enlightenment (there are a few around) may feel that it should. But, too bad, it doesn't. If modernity, then, doesn't necessarily lead to secularization (except in Sweden and in the faculty club of Delhi University), what does it lead to in the area of beliefs and values? The answer, we think, is clear: *It leads to plurality.*

WHAT IS PLURALITY, AND WHAT DOES IT MEAN FOR INDIVIDUALS AND SOCIETY?

By "plurality" we mean a situation in which diverse human groups (ethnic, religious, or however differentiated) live together under conditions of civic peace and in social interaction with each other. The process that leads to such a situation we would call "pluralization." Our thesis here, then, can be succinctly stated: *Modernity pluralizes.*

While that definition is simple, the empirical state of affairs to which it refers is highly complex. Before we attempt to explicate it, a terminological point: The situation we have called "plurality" is more commonly called "pluralism." We eschew this term because the suffix "ism" suggests an ideology rather than (as we intend here) an empirically available social reality. And it's as an ideology that the term "pluralism" first appeared. As far as we know, the term was coined in the 1920s by Horace Kallen, an American educator, and was intended to celebrate the diversity of American society. Think of the plurality/pluralism distinction this way: If "plurality" refers to a social reality (a reality that one may welcome or deplore), "pluralism" is the attitude, possibly expanded into a full-blown philosophy, that *welcomes* the reality. This terminological clarification helps to set off our thesis against the one we rejected earlier—the thesis proposing that modernity secularizes. Once again, secularity and the process of secularization which leads to it are concepts that refer to empirically researchable and (in this case) falsifiable social realities, which—like plurality—one can welcome or deplore. There's a long Enlightenment tradition, appropriately called "secularism," which welcomes secularization to the extent that it's believed to have taken place, and indulges in the hope that it will and should triumph in the future. We will have occasion to return to secularism later in our argument.

But back to our definition of plurality: The basic fact here is diversity in the groups making up a society. But our definition includes two further elements—civic peace and social interaction. These are important. One could, of course, have diversity without civic peace—the different groups engaged in violent conflict, possibly culminating in one group oppressing, enslaving, or even exterminating the others. It makes little sense to speak of plurality in that case. Alternatively, the diverse groups could exist side by side without interacting with each other—coexisting peacefully, if you will, but without speaking with each other. In that case too the distinctive pluralizing dynamic that's our focus here will not take place. An example of the first case—the antebellum South, with whites and blacks coexisting as slave owners and slaves. An example of the second case—traditional Hindu society, organized in castes that strictly avoid social interaction with each other (in accordance with prohibitions against commensality and connubium—eating with and marrying people outside the group—which are very effective in preventing interaction).

The reasons why modernity pluralizes are readily understandable: Through most of history most human beings lived in communities that were characterized by a very high degree of cognitive and normative consensus—that is, almost everyone shared the same assumptions about what the world is like and how one should behave in it. Of course, there were always marginal types, people who questioned these taken-for-granted assumptions—individuals such as, say, Socrates. But such individuals were quite rare. In other words, there wasn't much conversation between whatever diverse groups may have crossed each other's paths. The walls of social segregation were very high.

Modernity, with increasing speed and scope, weakens these walls. It has resulted in an ever-increasing proportion of the population living in cities, many of them huge—and cities have always been places where diverse groups go to rub shoulders on an ongoing basis. With that worldwide urbanization has come the spread of "urbanity"—an urban/urbane culture that's nurtured by plurality and in turn fosters the latter. Furthermore, there are massive movements of people across vast regions, again bringing very diverse groups into intimate contact with each other. Mass education means that more and more people are aware of different ideas, values, and ways of life. And, last but not least, modern means of mass communication—films, radio, television, telephones, and now the explosion of information through the computer revolution—have brought about an enormous increase in people's ability to access alternative approaches to reality. As a result of these processes—all endemic to modernity— plurality has reached a degree unique in history.

There have been plural situations in the past, of course. For centuries the cities along the Silk Road in central Asia enjoyed a true plurality, especially in the way different religious traditions interacted with and influenced each other—Christianity, Manichaeism, Zoroastrianism, Buddhism, and Confucianism. For longer or shorter periods, similarly plural situations prevailed in Moghul India, Hohenstaufen Sicily, and Muslim Andalusia (where the notion of *convivencia* was an early form of pluralistic ideology). Most significant for the history of Western civilization, the late Hellenistic-Roman period shows remarkable similarities with modern plurality, not least in terms of religious diversity. It's no accident that Christianity had its origins as a world religion in this particular milieu. But all these premodern cases of plurality were quite limited in scope. Take Alexandria in the Hellenistic period—probably as plural a society as one could find anywhere today (even in the absence of computers and cell phones). But if one took a boat only a short distance up the Nile, one would come upon villages that were as culturally homogeneous as any in the long history of Egypt—perhaps containing some people who didn't know that they were supposed to be part of the Roman Empire and perhaps some who had never heard of Alexandria. Even the most diligent organizers of eco-tourism today would have great difficulty finding comparable places of cultural homogeneity—and probably could find no cases of cultural isolation. (And, needless to say, if they succeeded, their own activity would rather quickly put an end to the pristine authenticity with which they sought to attract their clientele.)

What takes place under conditions of genuine plurality can be subsumed under a category used in the sociology of knowledge—"cognitive contamination." This is based on a very basic human trait: If people converse with each other over time, they begin to influence each other's thinking. As such "contamination" occurs, people find it more and more difficult to characterize the beliefs and values of the others as perverse, insane, or evil. Slowly but surely, the thought obtrudes that, maybe, these people have a point. With that thought, the previously taken-for-granted view of reality becomes shaky. There's ample evidence from social psychology that this process of mutual contamination occurs between individuals, even in experimental situations such as those pioneered by psychologists like Kurt Lewin or Milton Rokeach. Thus Lewin coined the term "group norm"—the consensus toward which any process of "group dynamics" tends. And Rokeach had a particularly curious example of this, in his classic study of *The Three Christs of Ypsilanti*—the latter location not referring to the Greek town in which Byron died, but to a mental hospital in Michigan. There were three patients in that hospital, each of whom thought he was Jesus Christ. One was too far gone in his psychotic isolation to be influenced by the other two. But Rokeach describes in fascinating detail how the other two, slightly less psychotic, came to terms with each other's Christological claims. Actually, they constructed an "ecumenical" theology to accommodate these claims.

The cross-contamination that happens on the level of individuals also happens between collectivities. In the history of religion, these collective processes of cognitive contamination are known as "syncretism." A classical case is the acquisition of the Greek pantheon by the Romans—Zeus becoming Jupiter, Aphrodite becoming Venus, and so on. The religious ideas (as well as other cognitive and normative categories) are "translated" from one worldview to another. Obviously, they don't remain unchanged in this process of translation.

In addition to igniting an explosion of cognitive contamination, over the past few centuries modernization has enabled science and technology to change the conditions of human beings fundamentally. This has been a vast transformation, affecting every aspect of a human society, and it can be described and analyzed in many different ways. For our purposes here, however, we want to focus on one aspect of that transformation—the gigantic shift *from fate to choice*.

HOW DOES MODERNITY'S SHIFT FROM FATE TO CHOICE AFFECT US?

This shift is clearly visible in the core of the process, the technological component. Imagine a Neolithic community dealing with a particular practical problem—say, how to make fire to heat the caves and to cook buffalo meat, the community's staple food. Imagine then the many centuries in which this community rubbed together two stones to produce the required ignition. That method was the only one they knew. A modern community obviously has a much wider choice among different sources of ignition and energy—different *tools*, if you will—not to mention different varieties of food. Indeed, a modern community has a choice not just among different tools, but among different systems of technology.

This expansion of the realm of choice affects not only the material aspects of human life, but also the cognitive and normative dimensions—the focus of our interest here. Take a fundamental question such as, What are men and women? In the aforementioned Neolithic community, there was hardly any choice in the matter: There was a clear and binding consensus about the nature of the two sexes, and about the norms following from this agreed-upon nature. Modernization has enormously increased the array of choice in this area of life. Typically, a modern individual can choose whom to marry, how and where to set up the household resulting from the marriage, what occupation to train for in order to support or help the household, how many children to have, and (last but not least) how to raise those children. Again, there are entire systems to choose from—systems of marital relations, systems of education, and so on. Additionally, the modern individual can select a specific personal identity, such as traditional or progressive, straight or gay, disciplinarian or permissive. In much of the developed world, modern identity is *chosen*, is a sort of project (often a lifelong one), undertaken by countless individuals.

This turns out to be a difficult proposition for many people. To help searchers decide who they want to be, a vast marketplace of lifestyles and identities has emerged, each actively promoted by advocates and entrepreneurs (two overlapping categories). A man in his mid-seventies, writing to his son to announce his fifth marriage, muses, "I'm finally discovering who I am" (and one can be sure that he will find a therapist or a support group that will confirm this assessment). Or a young woman with an Armenian name but no other Armenian connection, American-born and monolingual in English, when asked why she's taking a class in the Armenian language, replies, "Because I want to find out who I am" (and there can be no doubt that she will find an Armenian church or community organization that will welcome her in her new-found identity). It's in this context that Michael Novak, in his book *The Unmeltable Ethnic*, made the startling assertion that, in America, ethnicity has become a matter of choice.

In the mid-twentieth century, the German social philosopher Arnold Gehlen developed a very useful set of concepts to describe the aforementioned development. Every human society (including, presumably, a Neolithic one) allows some choices to its members, while other choices are preempted by taken-for-granted programs for action. The area of life in which choices are allowed Gehlen called the *foreground*, while the area in which choices are preempted he called the *background*. Both areas are anthropologically necessary. A society consisting of foreground only, with every issue a matter of individual choice, couldn't sustain itself for any length of time; it would lapse into chaos. In every human encounter people would have to reinvent the basic rules of interaction. In the area of relations between the sexes, for example, it would be as if Adam encountered Eve for the first time every single day and had to ask himself ever again, "What on earth should I do with her?" (and of course Eve would have to ask herself the corresponding question of what to do with him). This would clearly be an intolerable situation. Quite apart from anything else, nothing would ever get done; all available time would be occupied with inventing and reinventing the rules of engagement. Alternatively, a society consisting of background only wouldn't be a *human* society at all, but a collectivity of robots—a situation that, happily, is anthropologically (and presumably biologically) impossible.

The difference between foreground and background can be succinctly described as follows: Background behavior can be carried out automatically, without much reflection. The individual simply follows the programs laid out for him. By contrast, foreground behavior requires reflection—should one go this way or that? The balance between background and foreground has been greatly affected by modernization: The increase in choice that we mentioned earlier has led to a corresponding increase in reflection. Helmut Schelsky, a German sociologist born less than a decade after Gehlen, called this fact "permanent reflection" (*Dauerreflektion*). Permanent reflection can be seen today both on the individual and the societal levels. Individuals are led to ask constantly who they are and how they should live, and a vast array of therapeutic agencies stand ready to assist them in this formidable task. On the societal level the educational system, the media, and a multitude of (aptly named) "think tanks" ask the same questions about the entire society: Who are we? Where are we going? Where should we be going? It could be said, without much exaggeration, that modernity suffers from a surfeit of consciousness. No wonder that so many modern people are nervous and on edge.

A society's taken-for-granted programs of action are called "institutions," in Gehlen's lexicon. *Strong* institutions function as if they were instincts—individuals follow the institutional programs automatically, without having to stop and reflect. Just as without background a society would disintegrate into chaos, likewise without institutions (thus defined) no human society could survive. But the extent of institutional programs varies between societies, as does the relative size of foreground and background. When something is moved from foreground to background we can speak of "institutionalization," while the contrary process may be called "deinstitutionalization."

Take a simple illustration: A contemporary man gets up in the morning and has to make a number of sartorial decisions—to wear or not to wear a suit, to wear or not to wear a tie, and so on. These are foreground decisions; the relevant behavior has been deinstitutionalized. On the other hand, unless he's a very unusual individual or lives in a peculiar subculture, it won't occur to this individual to leave the house naked. Thus he's given a range of choices in terms of what to wear, but the fact that he'll wear *something* is still taken for granted—that is, continues to be robustly institutionalized. Obviously, however, such situations can change. Take an illustration from the area of gender relations: At some point in European history, norms of polite behavior came to include certain acts of courtesy extended by men toward women—say, letting women go through a door first. Clearly, this was not always so. (We can assume, for example, that this wouldn't have occurred to our aforementioned Neolithic types.) Perhaps it originated in the troubadour culture of the High Middle Ages. At that point—whenever it was—some man made a decision to let a particular woman go through the door first, perhaps even to hold it open for her and accompany the act with a little bow. After a time, in certain social milieus, this act became institutionalized. Thus, say, a hundred years or so ago, a middle-class European or American male would act unthinkingly according to the precept "Ladies first." Then came the feminist movement, and suddenly this whole area of intergender courtesy became deinstitutionalized. No longer could the dictum "Ladies first" be followed without reflection. The man now had to assess his female companion and then *decide* which course of action to pursue—hold the door for her (and perhaps get good marks for gentlemanly behavior, but perhaps be chided with an angrily snarled, "Thank you, but I'm not a cripple") or, in robust egalitarian fashion, go through the door first (with equally uncertain consequences).

Now, all this is pretty basic sociological theory. However, the point within the present argument is exceedingly important: *Modernity greatly enlarges the foreground as against the background*. Another way of saying this is: *Modernity tends to de-institutionalize*. Tellingly, Gehlen also called the latter process "subjectivization." Where previously the individual could go through life without too much reflection, by acting out the institutional programs, today the individual is thrown back on his or her *subjective* resources: What am I to believe? How should I act? Indeed, who am I? As noted earlier, to make this dilemma easier for people to handle, new institutions have emerged that offer entire packages of beliefs, norms, and identities to individuals. Gehlen called these "second-ary institutions." They do help to unburden the individual from the agony of too many choices, in that they allow a measure of unreflected behavior, but by their very nature they are weaker than premodern institutions. Because these secondary institution are likewise *chosen*, not given or taken for granted, the memory of the choice will persist in the in-dividual's mind—and with it the awareness, though dim, that sometime in the future this choice could be reversed and re-placed by a different choice. Jean-Paul Sartre proposed that "man is condemned to freedom." As a general anthropologi-cal proposition, this is rather questionable. But as a descrip-tion of *modern* humanity, it's remarkably apt.

HOW DOES PLURALITY AFFECT RELIGION, BOTH INDIVIDUALLY AND COLLECTIVELY?

Back to religion. The American language uses a reveal-ing term to describe an individual's religious affiliation—"religious *preference*." The term derives from the world of

consumer choices—one prefers this brand of breakfast cere-
als to that one. The term suggests choice: One doesn't *have* to
be Catholic; one *chooses* to be Catholic. But the term also suggests instability. Preferences can change: I may be Catholic
today, but tomorrow I might become an Episcopalian, or an
agnostic, or what have you. The (let's say) Californian version of the American language has an even more revealing
term: I'm *into* Buddhism. Tomorrow, of course, I may be *out
of* Buddhism and *into* Native American sweat lodges, and so
on. Naturally, even in the culturally and religiously volatile
American situation, most people don't swing from one such
preference to another every other day. They're held back
by the restraints of upbringing and family, and by the rather
pervasive desire to be consistent and to attain a measure of
stability. Yet the awareness that one *could* change one's preference is there all the time, and thus the possibility that at some
point one *might* do so.

To repeat: Modernization produces plurality. And plurality
increases the individual's ability to make choices between and
among worldviews. Where secularization theory went wrong
was in the assumption that these choices were likely to be *secular*. In fact, they may very well be religious. *Chosen* religion is
less stable (weaker, if you will) than taken-for-granted religion.
In addition, it *may* be more superficial (that is, have all the
triviality of consumer choices in a supermarket). But it need
not be. A passionate leap of faith such as Søren Kierkegaard
suggested is possible only in a situation in which religion is no
longer taken for granted. And that's anything but superficial.

The plural situation thus changes the place of religion
in the consciousness of individuals. One could describe
this consciousness as being layered in terms of "degree of

certainty"—from the "deep" level of taken-for-granted assumptions regarding the world (with "deep" not to be understood in Freudian terms—nothing "subconscious" about this
level of consciousness), through more or less stable beliefs,
"up" to the level of easily changed opinions: I'm an American
(and couldn't imagine being anything else). I'm politically
liberal. As of now, I'm inclined to prefer candidate X over
candidate Y. In individual consciousness, religion "percolates
up," as it were, from the deeper level of certainty toward the
much more fragile level of mere opinion, with various levels
in between those two. It's important to understand that this
change doesn't necessarily affect the *content* of religion. A traditional, taken-for-granted Catholic may adhere to the same
doctrines and practices as a modern individual who is Catholic by preference. But the location in consciousness of these
doctrines and practices will be different. Put differently, pluralization need not change the *what* of religion, but it's likely
to change the *how*. Again, to avoid facile generalization, we
should note that the voluntary character of religious affiliation inevitably means that individuals have a greater chance
of modifying the official doctrines or practices mandated
by their church—which means changing the *what* as well as
the *how* of religion. This is often expressed by a statement
such as, "I'm Catholic, *but* . . ." This qualifying *but* may mean
that the individual no longer believes in papal infallibility or
in the miracle of the mass, for example, or happily practices
methods of birth control not countenanced by the church.

But pluralization also changes the sociological character
of religious institutions and the relation of these institutions
to each other. Churches, whether they like this or not, cease
to be religious monopolies and instead become *voluntary*

associations. For some religious institutions (and their leadership), this is very difficult. In the Western world, the Roman Catholic Church is the most visible case of a religious institution being reluctantly but inexorably forced to operate as a voluntary association. Once the church can no longer rely on either cultural taken-for-grantedness or the coercive power of the state to fill its pews, it has no alternative but to try to persuade people to make use of its services. The relationship between church functionaries (in the Catholic case, hierarchy and clergy) and their lay clientele necessarily changes with that shift. Regardless of the theological self-understanding of the Catholic Church, its functionaries have to become more accommodating to the wishes of the laity, and the power of the latter will increase concomitantly.

Some Catholic observers have (pejoratively) called this process "Protestantization." It need not involve any doctrinal or liturgical concessions to Protestantism, however. It's simply an acceptance of the empirical fact that the church now is a voluntary association, dependent on the uncoerced allegiance of its lay members. This is "Protestant" only in the sense that much of Protestantism, and particularly the version that's become dominant in America, has operated in the form of voluntary association for a very long time. Put differently, Protestantism has had a comparative advantage in coping with the plural situation, but a religious body doesn't have to be Protestant to enjoy that advantage. The Catholic Church at first strongly rejected the idea that it's a voluntary association of believers, then willy-nilly accepted this status in countries where Catholics were in the minority (as in the United States) or where the state refused to continue its old supporting role (as in France).

In other words, sociology trumped ecclesiology. Afterward, the strong endorsement of religious liberty by the Second Vatican Council provided theological legitimation to the empirical process that had already happened. It's noteworthy that the two most influential Catholic thinkers formulating this legitimation came from the two mother countries of modern democracy—John Courtney Murray from the United States and Jacques Maritain from France. The Catholic case has been discussed here because it's the most dramatic. However, the same victory of the voluntary principle can be observed in other cases of erstwhile religious monopolies—in the Church of England, the Russian Orthodox Church, Orthodox Judaism, and, for that matter, Turkish Islam or Hinduism in many parts of India.

Pluralization also changes the relations of religious institutions with each other. They now find themselves as competitors in a free, or relatively free, market. Once they give up the project of restoring or creating anew a religious monopoly, they must somehow acknowledge their competitors. Richard Niebuhr (an American church historian, not to be confused with his better-known brother, the theologian Reinhold Niebuhr) coined a term to help address this phenomenon—the "denomination." Classically, historians and sociologists of religion have distinguished between two types of religious institution—the church, an inclusive body into which most members are *born*, and the sect, a small, exclusive body that individuals *join*. Niebuhr added the denomination, a distinctively American category. Its social character is closer to the church than the sect, but it's joined rather than born into by individuals, and it explicitly or implicitly recognizes the right of its competitors to exist. In other words, the denomination is the child of competition in a plural situation. Like all participating entities in a market, denominations must both compete and cooperate. The cooperation is expressed in innumerable "ecumenical" and "interfaith" activities, which civilize and to a degree regulate the competition. Needless to say, the denomination operates under the voluntary principle—the key empirical characteristic of the plural situation and the basic value ("the right to choose") of the ideology of pluralism.

The pluralizing dynamic, with all the aforementioned traits, operates most effectively under conditions of legally guaranteed religious liberty. However, even if a state tries to impose limits on religious liberty, the pluralizing dynamic has a way of intruding anyway. A good number of states in the contemporary world have tried to limit religious liberty (often under the banner of protecting their citizens from "proselytizing" missionaries)—most Muslim countries, but also Israel, Russia, India, and China. Some of these measures have been quite effective. But unless a truly totalitarian system is put in place, the pluralizing forces find their way in—through the huge channels of communication opened up by modern technology, through international travel both by citizens of the restrictive state and by foreign visitors to it, and often enough by missionaries who evade or defy state controls. It's very hard to stuff the pluralizing genie back into the bottle.

The foregoing discussion has centered on religion—on a distinctively modern form of polytheism. This pluralization constitutes a great challenge to all religious traditions, but modern societies have shown themselves capable of living with and adapting to it. The Enlightenment value of religious tolerance was undoubtedly fueled by revulsion against the atrocities of the wars of religion in Europe—and certainly no theological difference could justify the bloodbath of those wars. Pluralization, however, affects not only religion but also morality. And the pluralization of *values*, which are the foundation of morality, is more difficult to cope with than religious pluralization.

If I'm a Catholic who believes that transubstantiation occurs in the eucharist, and my neighbor is a Protestant with a much less miraculous view of this sacrament, it shouldn't be too difficult for the two of us to get along once we have achieved a certain degree of tolerance. After all, our theological difference in this matter need not affect any practical issue between us. We can peacefully agree to differ in this matter or in others like it. But what if we now face a plurality of moral norms? My neighbor believes in cannibalism, say, and practices this belief to the best of his ability.

The example of cannibalism is admittedly outlandish. But there are sharp moral differences within modern societies that lead to comparable conflicts. At the moment, the Muslim diaspora in Europe has presented Europe's democratic countries with sharp moral challenges: Can the liberal democratic state condone the honor killing of women who allegedly besmirch the honor of a family? Or can it condone genital mutilation? And what about the importation of underage wives, polygamously or otherwise? And what if their husbands claim the right to discipline them by beating them from time to time? There are less dramatic issues as well: Should European schools accept segregating boys and girls in school sports (or perhaps in other—or all—school activities), as Muslims request? Should women in head-to-

toe shrouds be permitted in government offices? Should old blasphemy laws be revived or reenacted, to proscribe insults to Islam? (Let it be remarked in passing that it may well be the case that these moral principles aren't authentically Islamic, but rather the result of cultural developments with nonreligious roots. The fact remains that these norms are legitimated in terms of religious liberty, and the host societies must decide to accept or reject this legitimation.)

One doesn't need to cross the Atlantic to find comparable challenges of moral plurality. The U.S. debates over abortion and same-sex marriage will suffice by way of examples. How does a society function if a sizable segment of its members believes that abortion is a woman's right, while another large segment believes that it's the murder of a child? And what if one sizable group in the society thinks that same-sex marriage is a basic civil right, while another group thinks that it's a revolting perversion? If I'm a Catholic, I can readily sit down amicably with my Protestant neighbor, without entering into an acrimonious debate over the nature of the eucharist. But can I have a friendly cup of coffee with a neighbor whom I consider to be a murderer or a pervert, or an advocate of murder or perversion?

Briefly put, moral pluralization today creates sharper challenges than religious pluralization. What's more, at least some moral judgments depend on a measure of certainty that one need not have in matters of religion. This is a problem that will have to be taken up later in this book.

2

THE DYNAMICS OF RELATIVIZATION

To reiterate: Modernity pluralizes. Modernity deinstitutionalizes. Another way of putting this: Modernity relativizes. In what follows, we intend to look in greater detail at the cognitive processes released by relativization.

WHAT IS RELATIVIZATION?

We can most easily see what relativization is by looking at its opposite: The opposite of "relative" is "absolute." In the realm of cognition, there are definitions of reality that have the status of absoluteness in consciousness. In other words, one is so certain of these that one can't seriously doubt them. The awareness of our eventual and inescapable death is probably the most pressing absolute. Philosopher Alfred Schutz called it the "fundamental anxiety." But this doesn't mean that the awareness of death is constantly present in consciousness. For most people, it's avoided by immersion in the everyday business (= "busy-ness") of living. For some, of course, the fear of death is mitigated by religious hopes for a happier hereafter.

Another absolute certainty is one's conviction that the outside world revealed by one's senses is real. This table in front of me, for example, is really there. I can't seriously doubt this. To be sure, in a philosophy class the instructor might suggest that I can't be sure the table is there if I'm not looking at it: *Close your eyes,* she says, *and turn around, now prove that the table is there.* Well, I can't prove it, so I'm forced to admit that I'm not really sure about the objective existence of the table outside my own consciousness. This class exercise is only a game, however. It's not serious. Even while I'm engaged in it, I know that the table is sitting out there all along. In other words, my knowledge of the outside world has the status of absoluteness.

Relativization is the process whereby the absolute status of something is weakened or, in the extreme case, obliterated. Although the evidence of one's senses carries with it a claim to absoluteness that's very hard to relativize, there's a whole world of definitions of reality that are *not* based on such immediate sense confirmation—the world of beliefs and values. The classical American sociologists Robert and Helen Lynd, in their studies of "Middletown" (which was their pseudonym for Muncie, Indiana), used the concept of "of course statements"—that is, statements about the world to which most people would respond by saying, "Of course." For example, "American democracy is superior to any other political system." The Lynds studied "Middletown" twice, in 1928 and 1937. At that time, most people in the town surveyed would have said, "Of course," when confronted with this statement about democracy (though who knows whether that would still be the case today).

Let's take a somewhat clearer case: An American today introduces a woman as his wife. He's asked, "Is this your only wife?" "Of course!" he replies, possibly in a tone of irritation. Now, obviously this little scene would play out differently in a country where polygamy is accepted and practiced. But even in America, the "of course" character of statements about one's spouse has shifted somewhat in recent years. Thus if a man mentions his "spouse" (rather than his "wife"), the question will obtrude as to whether this "spouse" is a woman or a man. To be sure, there are people in America who strenuously object to same-sex marriage. Even they, though, are aware of the fact that this alternative arrangement is now widely accepted and at least sporadically practiced (which is precisely why they're upset about it); and in consequence the previous "of course" status of heterosexual marriage has been put in question—that is, has been relativized. One could say that the project of the opponents of same-sex marriage is to reverse this relativization process and to restore the old absolute in law and in public consciousness.

The Lynds' term "of course statements" more or less corresponds to what Alfred Schutz called "the world-taken-for-granted." The latter consists of a typically large body of definitions of reality that are generally unquestioned. These are both cognitive and normative definitions—that is, assertions as to what the world *is* and as to what it *should be*. The taken-for-granted world is the result of a process of internalized institutionalization. In other words, the objective reality of an institution, such as marriage, now also has an objective status within consciousness. Marriage is "of course" out there—as husbands and wives play out their respective roles all over the place. But marriage is also "of course" in here—as I think about this particular institutional arrangement as being self-evidently the only plausible one. Now, institutions, because of their foundation in unreflected, habitual behavior, have a built-in quality of inertia. Unless sharply challenged, they have the tendency to persist over time. Such challenges mark the onset of relativization.

Each such challenge comes in the form of a shock: People are literally shaken out of their unreflected acceptance of a particular institution. These shocks can be collective or individual. Take an example from political institutions—say, a particular form of chiefdom in a tribal society. Collective shock: The tribe is conquered, the chiefdom is abolished, and a religious hierarchy is put in its place. In such a case, every member of the tribe will be shocked out of his or her taken-for-granted assumptions about the social order. Individual shock: An individual happens to encounter gross corruption or deceit in the house of the chief, and as a result the institution of chiefdom loses its taken-for-granted legitimacy. It's possible that this individual is alone in experiencing the shock; the rest of the tribe may happily continue to believe wholeheartedly in the institution of chiefdom, even if the incumbent chief leaves something to be desired. Furthermore, the shock and the ensuing relativization may be intended or unintended: The institution of chiefdom may be overthrown by a revolutionary conspiracy, or it may become almost unconsciously implausible as other forms of authority impinge on the tribe—say, the agents of a national government. And, as the previous example indicates, the relativizing shock can be sudden or gradual.

A group of sociologists revisited "Middletown" some twenty years ago. They found both continuities and changes. The single most notable change was that "Middletown" (and thus presumably American society as a whole) had become much more tolerant—of racial, ethnic, and religious minorities, and of people with alternative lifestyles. If you will, Middle America had become more cosmopolitan, more sophisticated. And it makes sense to explain this in terms of the aforementioned pluralization process. Since the 1930s, denizens of small towns in the heartland of the United States have had much more formal education (including, for a greater percentage, college). They have traveled much more—as tourists, as members of the armed forces—and had people from far away visit them or even settle among them (for instance, as resettled refugees). The interstate highway system has enormously increased their mobility, as has the quantum jump in automobile ownership. Moreover, they have been inundated with every sort of information by the recent explosion of communication media. As a result, communities such as "Middletown" have been yanked out of the comparative self-containment they enjoyed (or, if you prefer, suffered from) in the past. The mighty winds of a plural society have powerfully swept through their quiet, tree-lined streets.

Plurality, then, can lead to an increase in tolerance. What's more, tolerance increasingly occupies a very important place, sometimes the top place, in the socially approved hierarchy of values. This is particularly true in America, which, for well-known historical reasons, has been in the vanguard of plurality—and indeed of pluralism as an ideology. But tolerance as a primary virtue may now be found, minimally, throughout the developed societies of the West. Thus in Germany, Thomas Luckmann studied how moral judgments are made in everyday conversation, and he found that tolerance by far heads the list of values cited with approval. Conversely, lack of tolerance—narrowness, moral rigidity, being judgmental—is overwhelmingly condemned. There's good reason to think that one would come up with similar findings in other countries of western Europe.

In the history of American religion, there has been a steady expansion of the zone of tolerance. First, it was tolerance between and among all or most Protestant groups. Then Catholics and Jews came to be included. In 1955 Will Herberg published his influential book *Protestant—Catholic—Jew*, in which he argued very plausibly that there had emerged three authentic ways of being American, each under a broad religious label. In the half-century since then, Eastern Christian Orthodoxy has acquired legitimacy as another authentic American religious identity. (Can one now imagine a presidential inauguration without an Orthodox priest, tall black hat and all, appearing sometime doing the ceremony?) Indeed, there's now a growing acceptance of religious groups beyond Herberg's "Judaeo-Christian tradition." Americans increasingly see Islam as a genuine member of the Abrahamic family of faiths. This still leaves out the non-monotheistic religions—notably Hinduism and Buddhism, which are insisting that they be included in this American ecumenism. In western Europe, where there's a lesser degree of religious diversity, the development has been less dramatic, but the ideology of multiculturalism expresses a similar expansion of tolerance. (Unfortunately, it isn't *inevitable* that plurality will lead to ever greater tolerance, as history has shown us repeatedly. There can also be violent reactions against plurality, which we will look at later in this book.)

It's useful to distinguish between positive and negative tolerance. Positive tolerance is characterized by genuine respect and openness in the encounter with individuals and groups that hold values different from one's own. Negative tolerance is an expression of indifference: "Let them do their own thing"— "them" being those who believe or practice different things. The tolerance that has been growing in most of the developing world is largely of the second type. It has been elevated to a normative principle by the ideology of multiculturalism.

While the relativizing effects of plurality can be observed in large collectivities, even entire societies, it's important to understand that those effects are rooted in the micro-social interactions between individuals. And these in turn are rooted in a very basic fact about human beings—namely, that they're *social beings*, whose beliefs and values, whose very *identities*, are produced and maintained in interaction with others.

WHAT ARE COGNITIVE DEFENSES, AND WHY ARE THEY NECESSARY?

The notion of cognitive contamination, which was mentioned earlier, is rooted in this basic fact: As social beings, we are continuously influenced by those we converse with. That conversation will, more or less inevitably, change our view of reality. It's a given, then, that if we want to avoid such change, we'd better be very careful as to the people we talk with.

Psychologist Leon Festinger coined the very useful concept of "cognitive dissonance"—meaning information that
contradicts previously held views—or, more precisely, previously held views *in which we have a stake*. (It obviously doesn't
matter to us if new information contradicts previous opinions that are of little or no significance to us—say, about the
name of the capital of Papua New Guinea.) What Festinger
found out shouldn't surprise us: People try to avoid cognitive
dissonance. The only way to avoid it, however, is to avoid
the "carriers" of dissonance, both non-human and human.
Thus individuals who hold political position X will avoid
reading newspaper articles that tend to support position Y.
By the same token, these individuals will avoid conversation
with Y-ists but seek out X-ists as conversation partners.

When people have a strong personal investment in a
particular definition of reality—such as strong deeply held
religious or political positions, or convictions that relate directly to their way of life (that smoking is acceptable or even
fashionable, for example), they will go to great lengths to set
up both behavioral and cognitive defenses.

Behaviorally, as was pointed out earlier, this means avoid-
ing sources of dissonant information. But there are also cognitive defenses—exercises of the mind, if you will—to shore
up the favored view of things. Take defensive smokers. They
will search for material that challenges the dominant position that smoking is a health hazard—there are always dissidents, and the Internet now makes it quite easy to find them.
They will also look for ways to discredit the bearers of the
dominant view—they aren't qualified to assess the evidence
(because, for example, their degree is in the wrong field), or
they have a vested (and thus questionable) interest in taking

their position (they're on the payroll of an anti-smoking organization, perhaps).

Religious and political systems have been particularly adept at providing cognitive defenses for their adherents. A common defense strategy in that arena is to classify the bearers of dissonance under a category that totally discredits them and anything they might have to say—they're sinners or infidels, they belong to an inferior race, they're caught in false consciousness because of their class or gender, or they've simply failed to go through a particular initiation process leading to the putatively true view of things (such as conversion or proper ideological understanding). This strategy of discrediting the message by discrediting the messenger could be called "nihilation." In extreme cases, this can culminate in the physical "liquidation" of the inconvenient messenger.

Although cognitive defenses are common in the religious sphere, it's important to point out that they need not be religious or even ideological. The erection of such defenses is characteristic of any institution that seeks to dominate every aspect of the lives of its members. Sociologists have coined terms for this sort of institution: Erving Goffman wrote about the "total institution," while Lewis Coser spoke of the "greedy institution."

In the absence of some type of cognitive defense, the relativizing effect of conversation with "those others" will inevitably set in. What must happen then is a process of cognitive bargaining, leading to some type of cognitive compromise. What this means in the actual process of conversation between individuals was described in the classical psychological experiments by social psychologist Solomon

Asch in the 1950s. This is what happened in the simplest form of those experiments: A small group of individuals—usually students (the most convenient group for psychological experimentation on the part of professors)—was put in a room. All except one of them (the "victim," if you will) in each experimental group were instructed beforehand as to the nature of the experiment. When the group gathered, the "victim" was first shown an object—say, a piece of wood—and asked to estimate its length. He would come up with a reasonable estimate—say, ten inches. Then each of the other individuals in the group would be asked the same question and, as previously instructed, would come up with a grossly unreasonable estimate—twenty inches or thereabouts. The experimenter would then return to the "victim" and say something like this: "It seems that you have a very different opinion from everyone else here. Why don't you have another look at this object and see whether you might want to revise your original estimate." Well, almost every "victim" did so revise, typically saying something like: "I really don't think that it's twenty inches long. But maybe I did underestimate its true length. Maybe it's twelve or fifteen inches?" In other words, the "victim" was pulled toward what Kurt Lewin called the "group norm." Interestingly, the resistance against being so pulled increased sharply if there were two "victims" instead of one: The two dissidents would then huddle together and defend their (of course perfectly reasonable) original estimate. If "victims" and "non-victims" were evenly divided, a vigorous process of cognitive bargaining typically ensued, leading to a cognitive compromise—that is, a new "group norm." Asch apparently didn't modify his experiment to allow for status differences between the participants, but that would have made for an interesting twist. For example, we might assume that the outcome would be different if the "victim" was a professor and every other participant an undergraduate.

In Asch's experiment, the dispute was about a matter that could easily be resolved with recourse to a tape measure (though presumably the experimenter wouldn't have allowed this if any participant had suggested it), and yet the cognitive power of conversation nonetheless led to compromise. In other words, conversation had a relativizing effect even on the perception of a physical object, the length of which could, in principle, be decided by the application of a measuring device. When it comes to an individual's views about putative realities that *can't* be subjected to scrutiny by way of sense perception—such as religious or political views—*there are no commonly agreed upon measuring devices*. It follows that the power of conversation will be all the greater in such cases. What's plausible, and what's not plausible, will be largely determined by the nature of the conversation about it.

A useful concept of the sociology of knowledge to describe this phenomenon is that of the "plausibility structure." This is the social context within which any particular definition of reality is plausible. Religious institutions have always been aware of this. Take the classical Catholic formulation *Extra ecclesiam nulla salus*—"Outside the church there is no salvation." Translate this into sociological terminology: "No plausibility outside the appropriate plausibility structure." Think about it: It would be very difficult to maintain your Catholic identity if you were the only Catholic in a closed community of Tibetan Buddhists. It would help a lot if there were two of you; likewise, it would help a little if you could keep up a correspondence with Catholics on the outside. The best outcome (best in terms of your Catholic identity, that is) would be if you could escape your Tibetan captivity and return to a place where there were a lot of Catholics. Of course, the same concept pertains to the plausibility of non-religious beliefs and values. It's hard to be the only Marxist in town, or the only feminist, and so on.

Back to modernity and relativization: In premodern societies, plausibility structures are typically robust and stable. As modernity sets in, pluralization makes plausibility structures more fragile and temporary. We all become "victims" of Asch's experiment, now writ large to encompass entire societies. Therefore, we must be all the more careful regarding the people with whom we converse, as the Apostle Paul understood long ago when he admonished Christians not to be "yoked together with unbelievers" (2 Cor. 6:14). In the absence of such conversational caution, get ready for cognitive compromise!

HOW DOES RELATIVIZATION PLAY OUT IN TERMS OF RELIGION?

If you want to get a graphic sense of how relativization and its cognitive processes play out in terms of religion, we recommend a little bit of tourism. Go to Washington, D.C., take a car or a taxi, and go north on Sixteenth Street from near the White House toward Walter Reed Hospital. After a couple of blocks you will find yourself in a veritable orgy of religious plurality. Stretching north for miles, there's hardly a block without a religious building. There are churches of every Protestant denomination, including an African-American one. There's a large Roman Catholic parish. There's a Greek Orthodox and a Serbian Orthodox church. There are synagogues representing the three varieties of American Judaism. There's a Hindu center, a Buddhist center, a Baha'i center, and a big temple with an inscription in Vietnamese (presumably belonging to one of the syncretistic sects proliferating in that country). If one wants to include this in a list of religious edifices (an arguable point), there's also a large Masonic temple. There doesn't seem to be a Muslim presence, but the largest Washington mosque is only a couple of streets away.

It would probably not require much research to find out just why this particular stretch of Washington has been so fecund religiously. Perhaps it has something to do with zoning regulations. But the question that obtrudes here is this: Don't these people, in all these religious buildings, talk to each other? We don't mean formal interreligious dialogue—say, a conference on Judaism and Buddhism. We mean ordinary, friendly conversation—between the Serbian priest and the director of the Baha'i center, for example. We assume that there must be such conversations—if only because of the parking problems. Let's imagine that there's an important Baha'i event and use of the Serbian parking lot would be helpful. Presumably there's conversation about *that*. But does the conversation stop there? Possibly not. Before anyone really notices what's happening, the two interlocutors may find themselves deep in cognitive-contamination territory.

(Incidentally, on the topic of sociology-of-religion tourism, another suggestion could be made, this one about a stretch of roadway in Hawaii. Get a car and take the Pali Highway from Honolulu across to the other side of the island of Oahu. You'll find an array of religious offerings quite as plural as the one in Washington, only—as one would expect—with a stronger representation of Asian religions. It's not accidental that both locations are in the United States. To use a term coined by Talcott Parsons in another connection, America is the "vanguard society" of religious plurality.)

Whatever may be going on between Baha'i and Serbian Orthodox parking-lot negotiators in Washington, the major religious traditions *have* been talking with one another. This interfaith dialogue, while it has much older antecedents, surged forward after a major event in the late nineteenth century—the World Parliament of Religions in Chicago in 1893 (where, by the way, both Vedanta Hinduism and the Baha'i faith were first introduced to Americans). This dialogue has greatly increased in the last fifty years; it has become a minor industry occupying the attention of religiously interested intellectuals, and it has been formalized in departments of several major religious institutions (among them the Vatican and the World Council of Churches). Furthermore, there now exists a vast literature on this topic. We cannot provide an overview of this development here, but it would be worthwhile to describe briefly the major positions that have been taken by Christian participants in this interfaith conversation. (Note that the same categories could be used to describe positions taken by non-Christians.)

Three typical positions have been described—exclusivist, pluralist, and inclusivist. The exclusivist position concedes little if anything to the relativizing process: Christianity is reaffirmed in ringing tones as the absolute truth. As one would expect, this position is likely to be associated with doctrinal orthodoxy (be it Catholic or Protestant). It would be grossly unfair to equate this position with hostility or contempt toward adherents of other faiths, however. It's often accompanied by an attitude of respect for alternative religious traditions, even with a readiness to learn from them here and there. But there's no cognitive compromise with regard to a typically large number of doctrinal propositions deemed central to the faith.

In contrast, the pluralist position goes as far as possible in conceding to other traditions the status of truth, and in giving up any number of historical Christian doctrines in this process of cognitive bargaining. An important representative of the pluralist position has been the British Protestant theologian John Hick, an unusually prolific and articulate writer. Hick has hit on a very graphic metaphor: He calls for a "Copernican revolution" in our thinking about religion. Christians have traditionally thought of their faith as the center around which everything else in the world is circling. They now should think of their faith, advises Hick, as one of many planets circling around the sun of absolute truth— a truth that remains inaccessible to us in its fullness, which we can grasp only partially from the perspective of the one planet on which we happen to be sitting. While this is an attractive metaphor, Hick seems to exclude the possibility that some "planets" may not face the sun at all—in other words, he seems to imply that all "planetary" perspectives are equally valid—which is a hard argument to make, given the sharp contradictions between some of the perspectives. (Hick is aware of this problem. He tries to solve it by distinguishing between faiths in terms of their moral consequences: Religions are "true" as they produce "good" followers. However, if religion is to have any relationship with questions of truth, this is a very unconvincing criterion of differentiation. Take the analogy of science: The validity of the theory of relativity doesn't depend on Einstein's having been a good man.)

As one would expect, the inclusivist position is in the middle: It continues to affirm strongly the truth-claims of one tradition, but it's willing to go quite far in accepting possibilities of truth in other traditions, and it's willing to abandon elements of the affirmed tradition in making various cognitive compromises. If one takes this position, one must have some method of distinguishing what's central to one's faith from what's marginal (the latter falling under the category of *adiaphora*, or "things that don't make a difference"). Having made this distinction, of course, one will then be able to decide which elements of the tradition must be defended no matter what, and which elements may safely be let go.

Of course, the three positions are "ideal types" (to use a term of Max Weber's), with the lines between them not always clear. Although they can be useful, empirically each has certain dangers. The exclusivist may find himself in an antagonistic stance with regard to important aspects of contemporary culture—a difficult stance to maintain, given the pressures of relativization, and one that can suddenly collapse into a variety of relativism. The pluralist sooner or later must confront the fact that some of the "others" are so implausible or repulsive that they can't possibly be considered bearers of truth—which recognition may either plunge him into a nihilistic denial of any possibility of truth or, by way of reaction, throw him back into an exclusivist position.

As to the inclusivist, the danger here is that it's unclear just what's to be "included" and what's not, with a high degree of confusion in the offing.

While most people in the Western world tend toward an inclusivist position, all three positions of the official interfaith dialogue can also be found at the level of ordinary people talking about religion at work, across the fence separating two backyards, or in some other locale far removed from the conference rooms in which accredited representatives of various religious institutions engage in doctrinal discussions. These institutional discussions often have the character of boundary negotiations between nonexistent countries. Imagine, say, Catholic and Lutheran theologians engaging for several years in discussions of the doctrine of justification. Imagine further that, finally, they arrive at certain formulations that both can agree upon, and that they announce to the world that at least this particular doctrine need no longer divide the two churches. Well, one doesn't need to *imagine:* This exercise actually took place. One must ask, though, how many Catholic and Lutheran laypeople have the slightest knowledge of the official doctrines of their respective churches—or, for that matter, any serious interest in such doctrines. One must further ask whether the laboriously-arrived-at theological compromise has any relation to the lived piety of ordinary church members. And finally, one must suspect that there continue to be substantial differences between the two groups of laypeople—differences that were only papered over by the theologians.

Take a couple of, say, ten-year-old girls who live across the street from each other, play with each other, and sometimes talk about religion—usually because of some practice of one or the other. One girl is Catholic, the other Jewish. Perhaps the Catholic girl has heard that Jews consider themselves to be God's chosen people. Perhaps the Jewish girl is troubled by what someone told her—that Catholics think *their* church is the only one in possession of the full truth. What stances will they take in these discussions? Each may well take an exclusivist position—yes, this is what "we" believe, and while you and I can still be friends, these differences will stand. Alternatively, one or the other girl can be a pluralist: All roads lead to God; in the end Catholics and Jews are in the same boat of not knowing what the ultimate truth will turn out to be, and in the meantime we must be tolerant toward each other. If they both go the pluralist route, the Catholic girl will give up the notion of the one true church, and the Jewish girl will no longer adhere to the notion of a chosen people. Most likely, though, the two girls will end up as inclusivists: The Jewish girl will modify her previously held negative view of Jesus, without becoming a Christian; the Catholic girl will allow that truth can be found outside her church, also in Judaism, but she will continue to attend mass, pray the rosary, perhaps show great deference to the pope. Needless to say, ten-year-old girls are greatly influenced by the adults that are significant for them—their parents, teachers, clergy—as well as by the peer groups in which they move.

Both on the level of religious intellectuals and on the level of ordinary believers (whether ten-year-olds or adults), the inclusivist position is the most common, as noted above. Relatively few people convert to a totally different faith from the one they were raised in. The emotional pull of early socialization in a particular tradition is strong, and it's usually reinforced by the continuing influence of family and friends. Thus most people, at least in Western societies, engage in what Robert Wuthnow, an American sociologist of religion, has called "patchwork religion." That is, they patch together their individual religious "quilts," with bits and pieces coming from their original tradition as well as from other traditions. A French sociologist, Danièle Hervieu-Léger, has used the term *bricolage* to describe the same phenomenon on the other side of the Atlantic. (Her use of this term is quite different from its use in anthropology, going back to Claude Levi-Strauss, who coined it.) The term can be loosely translated as "tinkering"—as when one builds an edifice from different Lego pieces, sometimes resulting in very idiosyncratic constructions. Whatever other elements are added, the original religious tradition is still dominant. An individual may then be Catholic—recognizable as such by himself as well as by others—though, more precisely, he is "Catholic *but* . . ." The contents of the "but" may be quite far from what the church wants its members to believe. Thus, for example, European data show a surprisingly number of self-identified Catholics who believe in reincarnation. Both full-blown exclusivists and pluralists are rather thin on the ground.

The same tendency toward inclusivism can be observed in the area of morality. The so-called culture war in America is waged by committed activists on both the conservative and the progressive sides. To be sure, these activists have become very influential—indeed, they have become important parts of the constituencies of the two major political parties. With power hitters on both sides, the culture war is very real. But survey data indicate that most Americans are somewhere in the middle when it comes to the majority of the hot-button issues over which the conflict rages. Thus most Americans don't like abortion and want to limit it, but they don't want to make it illegal again. Thus most Americans disapprove of homosexuality, but they're willing to approve of civil unions by same-sex couples (as long as these aren't called "marriage"). And so on. In morality as in religion, most people shy away from complete relativism, but they're likewise leery of an absolutist affirmation of this or that value system or worldview.

WHAT IS MEANT BY "DIALECTIC OF RELATIVIZATION"?

Most of the above description could be read in such a way that the process of relativization would appear as a unilinear and inexorable process. So could our earlier discussion of modernization theory. As with modernization in general, however, its specific correlate of relativization is not unilinear or inexorable. Rather, relativization results in a dialectic by which it can, under certain circumstances, quickly mutate into a new form of absolutism. It's important to understand this dialectic.

Relativization, especially in its early phases, is commonly experienced as a great liberation. Much of European and American literature in the nineteenth and twentieth centuries described this experience in great detail. It typically involves a biographical trajectory from a narrow, provincial background into the broad horizons of modern life. Very often such a biography takes place in the context of urbanization: The hero or heroine moves into the big city from a village or small town. There occurs the great transformation of urbanity. Old prejudices and superstitions are discarded, and new ideas and ways of life are enthusiastically embraced. An essential part of this experience is that the individual is then confronted with a multitude of choices that previously were unimaginable—choices of career, intimate relationships, political and religious values, even the very definition of one's own identity. There's no reason to doubt that this process can be exhilarating. Very often it's just that.

Where is the dialectic here? It's quite simply this: The relativization that was at first experienced as a great unburdening now itself becomes a great burden. The individual now looks back with nostalgia to the lost absolutes of his past; alternatively, he or she searches for new absolutes. The liberation that is now sought is a liberation *from* the burden of relativity, *from* the many choices of the modern condition.

The nature of the burden is succinctly captured in an old American joke—admittedly not a very good joke, but it will serve here. Imagine a setting in one of the warmer states. Two friends meet. One says to the other, "You look depressed. Why are you so down? Are you still unemployed?" "No," says the second friend; "since last week I have a new job." "So what's the job?" "Well, it's in this orange grove. I sit in the shade, under a tree, and the pickers bring me oranges. I put the big ones in one basket, the little ones in another basket, and the in-between ones in a third basket. That's what I do all day. I sit in the shade under the tree and I put the oranges into these baskets." The friend says, "I don't understand. This sounds to me like a pretty soft job. Why does it depress you?" The response: *"All those decisions!"*

In German there's a telling phrase—*Qual der Wahl* ("agony

of choice")—that describes what the orange-sorter felt. Humans feel a reluctance to choose that is so profound it may be rooted in the biological constitution of the human species. Hence the need for what Gehlen called the "background" of institutions. Remember that institutions, if they function robustly, make it unnecessary for the individual to choose—at least in that segment of his or her life which is organized by the institutions in question. But this leaves a "foreground" in which choosing is unavoidable. As we've argued, modernity greatly expands this "foreground," *ipso facto* intensifying the agony of having to choose.

There are different responses to this situation. Probably the great majority of people deal with the burden of choice in the same way they deal with the unavoidability of death— by immersing themselves in the busy-ness of life, making the unavoidable choices with as little reflection as possible and muddling through as best as they can. (Not the worst-case scenario, one might think.) In addition, there's a typically small minority of people who *do* reflect and then come up with this or that mode of dealing with all those choices. What concerns us in this book is that group who are in quest of an authority that will declare an absolutely, ultimately correct choice.

Erich Fromm analyzed the psychology of totalitarianism in a book with the title *The Escape from Freedom*. While one might have some doubts about Fromm's psychoanalytic assumptions, the title is well chosen, and totalitarian movements *are* an attempt to escape the burden of freedom. Totalitarianism is indeed a kind of liberation. The individual, confused and terrified by "all those decisions," is presented with the comforting gift of renewed absolutes. We previously mentioned Jean-Paul Sartre's notion that man is condemned to freedom. We said that, while probably not universally valid as a description of the human condition, the statement aptly describes the condition of *modern* humanity. The new absolutism, then, is animated by a passionate refusal to submit to Sartre's "condemnation."

This is the great refusal of relativization. The proponents of the various versions of neo-absolutism have very seductive messages: "Do you feel lost in the 'patchwork' of religious possibilities? Here, surrender to the one true faith that we offer you, and you'll find yourself at peace with yourself and the world." Comparable messages are on offer to allay the vertigo of choices in morality, politics, lifestyles. And the message isn't lying: Fanatics *are* more at peace, less torn, than those who struggle daily with the challenges of relativity. This peace, however, comes with a price. Another Sartrean concept describes the price very well—the concept of "false consciousness" (*mauvaise foi*). Sartre describes this as the pretence that what one has in fact chosen is, on the contrary, a necessity imposing itself beyond one's own choosing. Sartre offers a raunchy illustration: A man is in the process of seducing a woman. He pushes his hand up her thigh. She realizes what he's up to and does nothing to stop him; she pretends that nothing is happening. In other words, although she's *choosing* to be seduced, she denies the choice by acting as if nothing is happening. That denial, says Sartre, is false consciousness. So, we would add, is the surrender to an absolute which is the core of every fanaticism: "I didn't *choose* this truth; it chose me; it imposed itself on me, and I can't resist it." Although this self-deception can work very well, there will always be the memory, however much repressed, that in fact this "truth" *was* chosen. If the memory is persistent, the new absolutism may collapse—and the dialectic then resumes. Thus in every relativist there's a fanatic waiting to come out in absolute certitude, and in every fanatic there's a relativist waiting to be liberated from any and all absolutes. We'll come back to this dialectic in our discussion of fundamentalism.

Much of this chapter has focused on religion. It's important to reiterate that secular cognitive and normative ideologies or worldviews are caught in the same dialectic of embracing and fleeing the results of relativization. The promise of absolute certainty can take many forms, of which religion is only one.

3

RELATIVISM

Relativization is quite simply a fact, the result of modern developments discussed in the preceding chapters. One may deplore the fact, or one may accommodate to it, or one may welcome it. Relativism had best be understood as describing the last of these three options: Relativity is embraced, accorded a positive normative status. The second option is an in-between one, between recognizing the fact and embracing it: There is as yet no welcoming it, but it is perceived as inevitable and one modifies one's behavior accordingly.

HOW DOES RELATIVIZATION AFFECT THE WAY PEOPLE SEE RELIGIOUS AND MORAL "OTHERS"?

The progressive responses of the Roman Catholic Church to the loss of its religious monopoly in the countries once called "Christendom" illustrate two of the three above-named options clearly. The Protestant Reformation of the sixteenth century was the most important challenge to the erstwhile monopoly—and Rome was in no way ready to accept this fact; rather, it tried as best it could to obstruct modernization (option one), by force if necessary. When the Reformation proved to be unquenchable, the church hierarchy had no choice but to accommodate to the relativization (option three)—not, initially, because their minds had been changed but because unalterable circumstances forced their hand. Among those circumstances, the wars of religion in Europe ended with the Peace of Westphalia, the Inquisition faded into a somewhat innocuous bureaucracy, and a note of reserved politeness entered communications with Protestants. Much later, in the 1960s, the Second Vatican Council, with its declaration on religious freedom, took an important step away from the old attitudes toward the Protestant "separated brethren," but even it didn't endorse relativism. Indeed, contemporary Catholicism still accommodates rather than embraces relativization. If there was any doubt about this, recent encyclicals reaffirming that the Roman Catholic Church is the only religious body in possession of the full truth of Christianity emphatically rejected the relativistic option.

The Catholic Church faced a comparable challenge in the rapid growth of Protestantism (mostly in its Pentecostal version) in Latin America, a part of the world once assumed to be solidly Catholic. There were some bishops who would have been quite happy to use robust methods to get rid of the challenge—in fact, there have been incidents of violence against Protestant churches and individuals. But Rome took a more gentle approach: Pope Benedict XVI, in his first visit to Latin America, spoke of Protestantism as a dangerous force (no ecumenical politeness there), but he didn't suggest any form of suppression; rather, he urged an intensive campaign of "evangelization" to win back those who had strayed from the Catholic flock. In other words, Catholicism recognizes the fact of relativization, and the behavior of the church has been modified accordingly, but relativization has not been accorded normative approval.

As was discussed in an earlier chapter, the pluralist position on interfaith relations, as exemplified by the work of John Hick—what we're calling "relativism" here—does indeed welcome relativity as an important new phase in the history of religion. Not only is the religious "other" accorded respect, and conceded freedom to believe and practice in ways different from one's own, but the "other" worldview is considered to be an instructive perspective on reality. In other words, the "other" is embraced as a harbinger of valid truth.

This kind of relativism isn't limited to religion. It applies to morality as well. In the relativistic view, there's no single, universally valid ethical system, but the moral values and behaviors of all, or virtually all, human cultures can be welcome additions to one's own ethical tradition. This type of relativism will always make exceptions, of course—cultures that, say, legitimize human sacrifice, or slavery, or treating women as inferior beings will not be considered valuable contributors to ethics. Every society must draw boundaries between acceptable and intolerable behavior.

This need for boundaries is becoming very clear in current discussions in Europe about the challenge of Islam. Some Muslim behaviors are widely deemed acceptable—such as individuals interrupting work to pray or women covering their hair with kerchiefs (though the acceptability of the latter is denied by some). Other behaviors are clearly unacceptable, such as "honor killings," female circumcision, and (in practice or even in theory) the death penalty for Muslims wanting to convert to another religion. But there are gray areas between these two poles, such as allowing parents to withdraw their daughters from participating in sports with boys, or making blasphemy a legal offense. In other words (to return to a question we raised earlier), what are the boundaries of the often invoked "European values"?

The final step in the celebration of relativity, going beyond a denial that truth is difficult to achieve, is to assert that the very notion of truth is meaningless and should be abandoned. It's not only difficult to escape the bias determined by one's location in history and society, extreme relativists argue; it's impossible and, in the final analysis, undesirable. There's no such thing as objective truth. Indeed, there are no objectively verifiable facts. There are different "narratives," all equally valid. This is the view taken by the so-called postmodernist theory, of which more later in this chapter.

The insight that beliefs about the world are relative in terms of one's location in space and time isn't new. It was expressed by both Herodotus and Ibn Khaldun, for example—themselves separated by space and time. In terms of the modern Western history of ideas, it was first proposed, pithily, by Pascal in the seventeenth century—in his statement that what is truth on one side of the Pyrenees may be error on the other. In Pascal's thought, this is related to another famous proposition of his, that of the wager of faith: Since one cannot be sure as to the truth of religion (precisely because of the relativity of all affirmations of truth), faith is a wager that, in a sense, one cannot lose. If religion is true, one's faith will be gloriously vindicated in the hereafter; if it's not true, and there is no hereafter, one will never know.

Modern Western philosophy began with René Descartes elevating doubt to a basic methodological principle in the seventeenth century. Given the fact that everything is subject to doubt, he said, what can one be certain of? Descartes answered that question by asserting that one thing which cannot be doubted is the self that is doubting. In other words, the quest for certainty is reduced to the supposedly indubitable reality of the subjective self: *Cogito ergo sum*. This has been called the great "turn to the subjective," and it dominated Western philosophy for several centuries.

At the same time, this very idea of the self was progressively undermined by the insight into its relativity: It isn't a universal idea. In archaic thought, the individual self is always embedded in the collective self of the clan or tribe: "I am what my tribe is." Or, to reverse the Cartesian dictum: "My tribe is, therefore I am." But even on the level of sophisticated theorizing, the alleged certainty of the self cannot be taken for granted. In the Upanishads, scriptures that are arguably the high point of Hindu thought, the individual self is finally identical with the innermost reality of the cosmos, the *atman* absorbed in the *brahman*. And, even more radically, one of the basic propositions of all schools of Buddhism is the denial of the reality of the self.

WHAT IS POSTMODERNISM, AND HOW DOES IT FIT INTO THE WESTERN WORLDVIEW?

In modern Western thought, the ever deepening sense of relativity, and therefore of the precariousness of all definitions of reality, is linked to three towering figures whose thought has had and continues to have an immense influence—Marx, Nietzsche, and Freud. Marx (who was mentioned in an earlier chapter as one of the founders of modern sociology) asserted that notions of truth are relative in terms of the individual's class position. Ideas are the "superstructure," which is determined by the "substructure" of the ongoing class struggle. Nietzsche saw relativity more generally in terms of the will to power. Ideas, he argued, are weapons in the struggle for power. Freud, on the other hand, saw ideas as rationalizations of subconscious cravings. All three thinkers developed what Nietzsche called the "art of mistrust," which is inevitably grounded in the insight that human ideas, including the idea of the self, are relative to an individual's social and psychological condition. This "art" was developed throughout the twentieth century by the sociology of knowledge, by post-Freudian psychology, and even by the discoveries in neurology of the way the brain functions. It was in the first years of that century that the philosopher Ernst Mach proposed that the idea of the self had become untenable (*unrettbar*—unsalvageable). But one can also say that the relativizers themselves are ongoingly relativized, and that in this process relativism is reduced to absurdity.

The most recent phase in this development of the "art of mistrust" has been so-called postmodernist theory. It has been associated mainly with two twentieth-century French thinkers, Michel Foucault and Jacques Derrida, though a somewhat milder form of the approach in the United States was proposed by Richard Rorty (who combined the theory with ideas derived from the American tradition of pragmatism, specifically as propagated by John Dewey).While both Foucault and Derrida acknowledged an indebtedness to Nietzsche, postmodernism has been alternately celebrated and attacked as a radical epistemological innovation. Given the history of ideas outlined above, it may be doubted that postmodernism constitutes a great epistemological breakthrough (though it has certainly engendered a novel and more-than-a-little-obscure vocabulary). Be this as it may, postmodernist theory does indeed represent a very radical restatement of the relativistic tradition.

There are differences among the three primary thinkers mentioned above (as well as Rorty, who has distanced himself somewhat from his earlier radical position). However, the main propositions of postmodern theory can be succinctly formulated as follows:

What passes for "knowledge" is always (as Nietzsche claimed) an instrument in a struggle for power. Assertions of disinterested knowledge are to be dismissed. There are no objective facts outside of power interests. The very notion of objectivity is an illusion and indeed is itself determined by specific power interests. For example, European scholars claimed to know, objectively, about societies and cultures of the Middle East, but this alleged knowledge was actually an instrument of imperialism and colonialism. (This example has been famously elaborated in a work by Edward Said, *Orientalism*, which has become a sort of icon of postmodernist theory.)

Although there are no objectively valid bodies of knowledge, there are different "discourses," always in the service of some existing or aspired-to power structure. Each discourse is a body of "narratives." Thus there is a colonialist and an anticolonialist narrative. As power structures change, the earlier narrative must be changed. For example, Japanese nationalism fostered a narrative of heroic militarism; after the defeat of Japan in World War II, and the assumption of power by a democratic regime in that country, a narrative of Japan as a peaceful nation appeared. The old and the new narratives have been competing ever since, and the debate over which is the correct narrative of Japanese history has been prominent in domestic politics (such as the arguments over history textbooks for schools) and foreign relations (such as the strenuous objections by China and South Korea to the way Japanese imperialism has been presented in those textbooks). It is illusionary to try to decide which narrative has greater validity. There is no hierarchy of truth as between the different narratives. Because all narratives are equally valid, at least in principle, it's futile to debate which narrative is closer to the truth. Instead, one must "deconstruct" all narratives, which means to show up their grounding in this or that power interest. This is relativity carried to an extreme degree. The most radical postmodernists (notably Derrida) have proposed that one must give up the entire project of finding the truth through reason and empirical science. (Derrida calls this project "logocentrism.") To the extent that this project is central to modernity, the proposal now is to give up modernity. Differently put, one must repudiate the Enlightenment.

HOW DO RELATIVISTS AVOID
RELATIVIZING THEMSELVES?

All versions of relativism have one problem in common: How are relativistic thinkers to exempt their own thought from relativistic "deconstruction"? After all, each relativist has a particular location in space and time, which location, if that individual's approach is correct, must determine his or her thought as much as anyone else's. Put differently, the relativist's thought is but one "narrative" among many others that are just as valid. To solve this problem, different versions of relativism have proposed what's sometimes called an "epistemological elite"—a select group of people who are assumed to be immune to the ravages of relativization. This elite has the sole custody of truth; everyone else "just doesn't get it." Needless to say, relativistic thinkers must assert that they are members of the elite, fellow custodians of truth.

The history of Marxism very clearly (and quite amusingly) illustrates this problem—how to exempt relativistic thought from relativistic deconstruction—and various attempts to solve it. Marx's thought is highly relativistic: Ideas are not to be understood on their own terms, but as ideological expressions of class interests. Those who don't understand them in this way (such as those who hold ideas different from Marx's) are in "false consciousness." Who, then, has the *correct* consciousness? In a rather involuted argument, Marx assigns this consciousness to the proletariat, which by virtue of its oppressed condition is free from the distortions of ideology. In other words, the proletariat is the epistemological elite. The proletariat, imbued with correct consciousness, is thereby empowered to become the bearer of revolution. It is left unclear how Marx—an individual of impeccably bourgeois origins, who was financially supported through most of his career by Friedrich Engels, a successful capitalist—manages to be a member of the aforementioned epistemological elite. Presumably one can be a sort of honorary proletarian (or, if you will, a proletarian by adoption). Georg Lukacs (a writer much esteemed by neo-Marxists, and very much a scion of the Hungarian bourgeoisie) is another telling example.

The subsequent history of Marxism shows various attempts to deal with a very inconvenient fact: The proletariat failed to develop the consciousness that Marxist theory had assigned to it. Instead of becoming the bearer of revolution, the working class in the developed capitalist countries gave its allegiance to social-democratic parties and/or labor unions, which sought to reform the capitalist system rather than mobilizing to overthrow it. The first successful Marxist revolution took place in Russia, a country with a very small working class. This revolution was led by bourgeois intellectuals, such as Lenin and Trotsky, and recruited its troops from the peasantry and from marginal groups (the ones that Marx called the *Lumpenproletariat*, which often overlapped with the criminal underworld). The inconvenient fact of the recalcitrant proletariat led to a simple imperative: The epistemological elite had to be located elsewhere.

Lenin located the elite in the Communist Party, which was supposed to be the "vanguard of the proletariat." If the working class couldn't or wouldn't fulfill its revolutionary mission, the party would do so on its behalf. This reformulation was very useful politically, of course: The party became the sole repository of truth; whatever was outside the party line (even if that line changed dramatically) was *ipso facto* false consciousness. The overriding maxim was simple: The party is always right. The intellectual contortions to which this maxim would lead are amply documented in the literature of the Stalin era, which carried the Leninist concept of the party to its logical and massively murderous conclusion. A classical example of this literature is Arthur Koestler's novel *Darkness at Noon*. The action of the novel takes place during the bloody purges of the 1930s. The main character, a committed Communist, confesses to a long list of crimes against the Soviet state, knowing full well that all the charges against him are false and that his execution is inevitable. He confesses to these fictitious crimes at one of the show trials so common during this period. And he does this not because he was tortured, or because his family was threatened, or because he was promised leniency; no, he does it because his interrogator convinces him that this was the last service he could render to the party that he had faithfully served all his life! There are, of course, religious parallels to this type of ideological masochism: The infallible party has simply taken the place of the infallible church. Recall the statement by Ignatius of Loyola to the effect that, if he perceived something as white but the church pronounced it to be black, he would cheerfully assert its blackness.

Not surprisingly, Marxists of a more humanistic bent were repelled by Lenin's self-serving concept of the party. Rosa Luxemburg, the German Communist agitator who was murdered by right-wing nationalists shortly after World War I, found another solution to the problem of the epistemological elite: The elite was supposed to be what she called the "colonial peoples"—much later to be baptized as the "Third World." In this view, the entire developed world, including its reformist working class, constitutes the "bourgeoisie." The class struggle is between it and the reinvented "proletariat" of the colonized.

Luxemburg's ideas didn't go very far at the time, but they got a new lease on life in the "Third Worldism" of the post–World War II era. Western Marxists (and indeed leftist or countercultural Westerners in general) went in search of truth in revolutionary movements in Latin America, Africa, and Asia. "Marxisant" liberation theologians convinced Latin American bishops to proclaim that the church had to have a "preferential option for the poor." This preference, while generally construed in political terms—"The church should always align itself with the aspirations of the poor"—was sometimes applied epistemologically—"The poor are always right." This latter maxim led to some remarkable intellectual distortions: Various bloody and corrupt movements in developing societies were enthusiastically embraced because they purported to act on behalf of the poor. It didn't matter that the poor hadn't elected these movements to speak for them—any more than the working classes of Europe and North America had elected the Communist Party to be their vanguard.

Antonio Gramsci was probably the most appealing of the Marxist humanists. He wrote many of his works in prison, where he spent a long time under the Fascist regime in Italy. His solution to the problem of relativizers being relativized was even more imaginative than Lenin's or Luxemburg's. It was also embarrassingly self-serving: The epistemological elite was to be, of all things, the intelligentsia. To reach this conclusion, Gramsci had to modify the substructure/superstructure model of mainline Marxism. The traditional view understood the superstructure (which includes the world of ideas, in addition to everything else implied by the notion of culture) as directly determined by the substructure (essentially the class system and its conflicts). Lenin went so far as to call the superstructure a direct "reflection" of the substructure. Gramsci rejected this determinism, arguing that the superstructure has a dynamic of its own and can in turn act back upon the substructure. And, of course, the bearers of this dynamic are the intellectuals. Not surprisingly, this view became very popular in intellectual circles and especially among students. In the upheavals of the late 1960s, in Europe and in the United States, many of the rebelling students understood their social role in essentially Gramscian terms: They were the truly authentic revolutionaries. A similar conclusion was arrived at by Karl Mannheim, though by a different route. Mannheim, a Hungarian-born scholar, brought the so-called sociology of knowledge into the English-speaking world. He insisted that all human knowledge (with the possible exception of pure mathematics) was determined by its social context—that is, was historically and sociologically relative. But, he claimed, one group is exempt from this determinism—the intelligentsia. Mannheim called this group the "free-floating intellectuals" (*freischwebende Intelligenz*). Their immunity to relativization was due, in Mannheim's view, to the absence of class interests—an absence that gave the intelligentsia the freedom to look at reality without false consciousness. Mannheim realized that intellectuals don't always perform this role, but he posited it as a normative possibility.

Nietzsche located the epistemological elite in the mystical ideal of the "superman," who transcended vulgar interests by his ascetic purity of thought. Nietzsche was no sociologist, so the social location of this superior specimen was left unspecified and should probably be understood as something yet in the future. This notion was taken up by the Nazis, who defined the superiority of the "superman" in racial terms, but Nietzsche can't be held responsible for this utilization of his idea.

Freud, another great relativizer, saw psychoanalysis as the path through the false consciousness of repression and rationalization. The epistemological elite, when seen in Freudian terms, is the community of the psychoanalyzed. This perspective offers a highly efficient method for disposing of cognitive dissonance: Anyone who disputes the findings of psychoanalysis is "resisting" its uncomfortable truths; only those who've been through the ordeal of the analysis can understand these truths—the others "just don't get it."

And postmodernist theory? Presumably, for postmodernists the elite consists of those who have mastered the arcane jargon of this school of thought. Any young academic aspiring to tenure in, say, a department of literature will immediately understand this.

Dissemination of the formula "You just don't get it"—unless, that is, you're one of us—has extended far beyond formal schools of this or that thought. There's a feminist ideology, for example, which asserts that you have to be a woman to understand the truth of women's oppression. Likewise, there's a "black consciousness" ideology which asserts that only African-Americans can understand African-American concerns. Substitute gender, sexual preference, race, ethnicity, or any other collective identity—and you will encounter the aforementioned relativist formula.

WHAT IS WRONG WITH RELATIVISM?

The most important thing that's wrong with relativism is a false epistemology. Put simply, all versions of relativism exaggerate the difficulties of ascertaining truth—at least to the extent that truth can be empirically sought. There are *facts* in this world, and in seeking to ascertain facts, *objectivity* is possible.

The facticity of physical reality is self-evident to anyone who isn't caught in philosophical abstractions. In a famous episode from the history of philosophy, Bishop Berkeley explained to Dr. Johnson that there's no way of disproving the proposition that the outside world is nothing but a figment of one's imagination. Evidently this conversation occurred while the two men were on a walk. Dr. Johnson kicked a stone across the road and exclaimed, "Thus I disprove it!" The physical fact of a stone lying on the road can be ascertained by any sane individual—regardless of class, race, or gender.

There are also social facts that can be proved. One of the foundational statements in the development of sociology was Emile Durkheim's instruction to "consider social facts as *things*." And what is a thing? Whatever resists our wishes, imposes itself on us whether we like it or not. This thing-like quality pertains to all functioning institutions, beginning with language. Thus a person learning a foreign language may protest that its grammar is illogical, that its pronunciation is awkward, and so on. The language teacher may reply, "Sorry, but this is the grammar and the pronunciation of this particular language—and you'd better learn them if you want to be understood." Now, we know that language as well as other institutions can be changed, and sometimes changed deliberately. But as long as institutions are firmly established in a society, they partake of this quality of facticity. Take the law. Again, one may opine that it's illogical, unclear, immoral. A lawyer can respond much as the teacher did about grammar: "I'm sorry, but whether you like it or not, this happens to be the law that applies to your case."

As mentioned earlier, a term much used by postmodernists is "narratives." These are supposed to be beyond testing by an allegedly illusionary reality "out there." Leopold von Ranke, a nineteenth-century German historian, defined the science of history as the effort to understand "what really happened" (*"wie es wirklich geschehen ist"*). Postmodernists reject this idea as being illusionary, and possibly undesirable as well. There are no facts, according to postmodernists—only narratives, all of which are epistemologically equal (though, as we have seen, some narratives are epistemologically privileged—those of the proletariat, the psychoanalyzed, and so on).

To better understand the nature of fact and narrative, let's look at an event that's very relevant to the relationship between Japan and China at the time of this writing—the so-called rape of Nanking. A number of historians have tried hard to determine just what occurred. There's overwhelming evidence that, after capturing what had been the capital of China, Japanese troops engaged in an orgy of murder, rape, and pillage. Thousands of civilians were killed. Can one deny these facts by assigning them to narratives? If there's a Japanese narrative and a Chinese narrative, is it futile to ask which is closer to the facts? The same question can be asked about the Holocaust: Are there *facts*, or is there only a Nazi narrative to be placed alongside a Jewish narrative? All that the postmodernist can do is to "deconstruct" each narrative down to the power interests it legitimates. (Come to think of it, a premature postmodernist was the late Nazi propaganda minister Joseph Goebbels, who declared, "Truth is what serves the German people.")

Admittedly, it's often difficult to arrive at an objective account of the facts. The interests and prejudices of any observer get in the way. But that's no reason to give up on the effort to be objective, which effort is definitely enhanced by an awareness of one's own interests and prejudices. There's a simple test as to whether the effort has been successful: If an observer is compelled by the evidence to make statements about facts that are contrary to his interests or prejudices, that observer is likely being objective.

Now, these considerations apply to facts that can be empirically ascertained—say, by a physicist, historian, or social scientist. But there are also truth-claims of a moral or religious sort, which are *not* susceptible to empirical exploration. The historian can't decide whether slavery was morally reprehensible; the social scientist can neither verify nor disprove the existence of God. However, here too reason can be used to make judgments on the plausibility of this or that morality, this or that religion. Later in this book we shall come back to this matter.

In the social sciences, the term "constructivism" has gained currency as denoting a postmodernist approach: There are no objective facts, only interest-driven "constructions." The term, in all probability, was originally meant to allude to the book *The Social Construction of Reality*, by Peter Berger and Thomas Luckmann (1966). Just as Marx allegedly said,

"I am not a Marxist," Berger and Luckmann have repeatedly announced, "We are not constructivists." The comparison between postmodernist theory and the Berger/Luckmann reformulation of the sociology of knowledge is useful for the clarification of postmodernism. Perhaps the word "construction" in the Berger/Luckmann volume was unfortunate, as it suggests a creation *ex nihilo*—as if one said, "There is nothing here but our constructions." But this was not the authors' intention; they were far too much influenced by Durkheim to subscribe to such a view. What they proposed was that all reality is subject to socially derived *interpretations*. What much of postmodernist theory proposes is that all interpretations are equally valid—which, of course, would spell the end of any scientific approach to human history and society. And some postmodern theorists have maintained that nothing exists except or outside these interpretations—which comes close to the clinical definition of schizophrenia, a condition in which one is unable to distinguish reality from one's fantasies. Put simply, there's a world of difference between postmodernism and any sociology of knowledge that understands itself as an empirical science.

All forms of relativism contradict the commonsensical experience of ordinary life (which is precisely what Dr. Johnson had in mind when he kicked the stone). Common sense is cognizant of an outside reality that resists our wishes and that can be objectively accessed by reasonable procedures. Even a postmodernist theorist operates on this assumption in ordinary life. Suppose that a postmodernist consults his doctor. He wants to know whether a tumor is benign or malignant. He expects the doctor to give an answer based on objective methods of diagnosis and to do so irrespective of personal feelings about the patient. Or suppose that a student has submitted a term paper. She expects the instructor to grade the paper "fairly"—that is, objectively and apart from any personal feelings. The student, whatever her theoretical inclinations, will vigorously protest if the instructor returns the paper with a failing grade and a note saying, "I hate your guts, and that's why I'm failing you." (A student is unlikely to protest if the note says, "I like you, and therefore I'm giving you an A"—but the principle is the same.) There's something wrong with a theory that contradicts the self-evident experience of ordinary life. The purpose of theory is to *elucidate* experience, not to deny it.

The foregoing discussion might lead a reader to believe that relativism is mainly a pastime of theorists. That would be a serious misunderstanding. Rather, relativism has massively invaded everyday life, especially in Western societies. This isn't because relativistic theories have gained many converts (though this is probably the case for the increasing number of people going through an educational system in which teachers propagate relativistic ideas). Rather, popular relativism is the result of more and more people experiencing the far-reaching effects of the plurality discussed in an earlier chapter of this book. History isn't an extended theoretical seminar. It's shaped by the living experience of large numbers of people, most of them without knowledge of or interest in the theories debated by intellectuals. Thus relativism is spread, not primarily through the propaganda of intellectuals, but by numerous conversations at places of work, across backyard fences, and even by children of different backgrounds meeting each other in kindergarten.

We have argued that relativism is epistemologically flawed.

It's also politically dangerous. Once again, we can refer to a core insight of Emile Durkheim: A society cannot hold together without some common values (which he called the "collective conscience" of the society). Without those shared values, a society begins to disintegrate because the behavioral choices of individuals become completely arbitrary. Morality becomes a matter of idiosyncratic preference and ceases to be subject to public argument: "You think that slavery is okay; I don't. You have the right to your opinion. I won't be judgmental. I won't try to impose my views on you." This, not so incidentally, has been a recurring position taken by American politicians: "I think that abortion is homicide—but I won't try to impose my views." This position is not only intellectually incoherent (and almost certainly insincere); it implies that public life should have no relation to morality. This implication will be denied by the aforementioned politicians, of course, but it's there all the same.

Relativism, with its individual rather than collective morality, is an invitation to nihilism. It can also be described as decadence—defined as a situation in which the norms that hold a society together have been hollowed out, have become illusionary and likely risible, and (most important) have undermined the trust that other people will behave in accordance with collectively shared norms. A decadent society doesn't have much of a future: It lacks the will to defend itself even against very real dangers to its very existence.

4

FUNDAMENTALISM

The term "fundamentalism" has been used very loosely in recent discourse—in academia, in the media, and, indeed, in ordinary language. Thus Muslim suicide bombers, Evangelical missionaries, and observant Orthodox Jews have all been called "fundamentalists"—a broad usage that induces serious flaws in perception. Sometimes it seems as if any passionate religious commitment is deemed to be "fundamentalist." In view of this, it would help to pay attention to the origin of the term, which came out of a very specific milieu of American Protestantism.

HOW DID THE TERM "FUNDAMENTALISM" ORIGINATE?

In the early twentieth century, two wealthy Los Angeles laymen set up a fund of $250,000 (a formidable amount at that time) to finance the production and distribution of a series of tracts defending conservative Protestantism against the inroads of liberal, modernizing theology. It's important to understand that even then this exercise was *reactive*—a reaction against trends deemed to be a threat against religious truth. The series was called "The Fundamentals." Beginning in 1910, altogether twelve booklets were published and widely distributed. By the time the twelfth volume came out just before World War I, three million booklets had been distributed, launching what came to be called the fundamentalist movement in English-speaking Protestantism.

The movement was both ecumenical and international. Although the promulgated orthodoxy was essentially Reformed (then represented most forcefully by Princeton Theological Seminary), the authors included prominent Presbyterians, Anglicans, and Baptists from both the United States and Britain. Naturally, there were significant differences among the authors. Despite these differences, however, there were a number of common themes defining the movement—an insistence on the unique authority of the Bible and on the supernatural events it recounted, a belief in conversion and a personal relationship with Jesus Christ, and a strict moral code. These themes continue to be central to the broad and quite diverse community called "Evangelical" in the United States and Britain, with the term "fundamentalist" generally rejected by most members of this community. The designation of "fundamentalism" is rather doubtful even in its original American Protestant context, papering over significant differences. It becomes even more doubtful when applied to Muslims or Jews, let alone Hindus or Buddhists. To make the usage more doubtful still, there are also *secular* "fundamentalists," representing a great diversity of ideological commitments and frequently exhibiting a passionate militancy that resembles that of certain religious movements.

This is a problem that's very familiar to social scientists: Terms become fuzzy if they're widely used, both in the vernacular and in academic discourse. People can deal with this in one of two ways. Individuals or groups can eschew such terms completely and create their own, new, sharply defined terminology. (Such terminologies are typically barbaric in what they do to ordinary language; worse yet, they make the writings of social scientists incomprehensible to the uninitiated—a kind of secret language.) The other alternative is to accept the terms as commonly used, but to sharpen them in order to better understand the social reality to which they refer. This is the tactic we prefer.

WHAT ARE THE CHARACTERISTICS OF CONTEMPORARY FUNDAMENTALISM?

The term "fundamentalism," as conventionally used today, refers to an empirically ascertainable reality. We would emphasize three aspects of this reality:

As in the prototypical American case, *fundamentalism is a reactive phenomenon*. In other words, it isn't a timeless component of this or that tradition. The reaction is always against a perceived threat to a community that embodies certain values (religious or secular). In the contemporary situation, the reaction is, precisely, against the relativizing effect of modernity discussed earlier in this book.

It follows that *fundamentalism is a modern phenomenon*. This point has been frequently made with regard to the efficient use of modern means of communication by fundamentalist movements. Fair enough. But fundamentalism is modern in a more profound sense. It can be understood only against the background of modernizing and relativizing process. Another way of stating this second characteristic of fundamentalism is that, despite its common claim to be conservative, to go back to the alleged golden age of a particular tradition, *fundamentalism is very different from traditionalism*. The difference can be simply put: Traditionalism means that the tradition is taken for granted; fundamentalism arises when the taken-for-grantedness has been challenged or lost outright.

An episode from the nineteenth century will serve by way of explication: Napoleon III, accompanied by the Empress Eugenie, was on a state visit to England. Eugenie (whose earlier history was, to put it mildly, not exactly aristocratic) was taken to the opera by Queen Victoria. Both women were quite young and regal in their demeanor. Eugenie, the guest, went into the royal box first. She gracefully acknowledged the applause of the public, gracefully looked behind her to her chair, and then gracefully sat down on it. Victoria was no less graceful in her demeanor, but with one interesting difference: She didn't look behind her—*she knew that the chair would be there*. A person truly rooted in a tradition takes the "chair" for granted and can sit on it without reflection. A fundamentalist, on the other hand, no longer assumes that the "chair" will be there; he or she must *insist* on its being there, and this presupposes both reflection and decision. It follows that a traditionalist can afford to be relaxed about his or her worldview and quite tolerant toward those who don't share it—after all, they're poor slobs who deny the obvious. For the fundamentalist, these others are a serious threat to hard-won certainty; they must be converted, or segregated, or in the extreme case expelled or "liquidated."

The final characteristic builds on the first two: *Fundamentalism is an attempt to restore the taken-for-grantedness of a tradition, typically understood as a return to a (real or imagined) pristine past of the tradition.* Given what was proposed in the preceding paragraphs, this understanding is seen to be illusionary. The pristine condition *can't* be regained, and therefore the fundamentalist project is inherently fragile. It must continuously be defended, propped up. This is often done in tones of aggressive certitude. And yet, as we noted earlier, however much the fundamentalist may repress the memory that his or her position was *chosen*, that memory remains, along with the knowledge that any choice is, in principle, reversible.

HOW DO FUNDAMENTALISM AND RELATIVISM COMPARE?

If the above considerations are empirically valid, it's clear that relativism and fundamentalism are two sides of the same coin. Both are profoundly modern phenomena, and both are reactions to the relativizing dynamic of modernity. The relativist *embraces* the dynamic; the fundamentalist *rejects* it. But the two have much more in common than either one would have with a genuine traditionalist. Their commonalities explain why we said, at the end of the second chapter, that in every fundamentalist there's a relativist waiting to be liberated, and in every relativist there's a fundamentalist waiting to be reborn.

We introduced the distinction between *traditionalism* and the fundamentalist project of *restoring tradition* with the example of Empress Eugenie and Queen Victoria. Let's look at another illustration of that difference. In the 1970s, an American social scientist went to Tanzania to study a project that was attempting to create a genuinely African version of socialism. An institutional instrument had been developed by the Tanzanian government to carry out this policy— the so-called Ujamaa villages (*ujamaa* being a Swahili word denoting solidarity). These villages were indeed socialist, in that there was no private ownership of land; in fact, they were somewhat similar to Israeli kibbutzim. At the time of the social scientist's visit, people joined the Ujamaa villages voluntarily (or so it was claimed), though peasants were later forced into these villages.

One important characteristic from the beginning was that people from different ethnic and tribal groups would come together to inhabit these villages, sharing their various cultures. As the guide explained to the visitor, in order to foster *ujamaa* among the diverse population, certain times were set aside for groups to perform their traditional dances. Later, reflecting on this, the visitor engaged in a mental experiment: He imagined that there were two films of these dances, one set in a traditional village and one in the village he had just visited. He imagined further that the two films were identical in what they depicted—the same dances, the same drums and chants, perhaps even the same dancers. *Yet the two events would be totally different.* In a traditional village, people danced at times established by the tradition, they danced without a reflected-upon utilitarian purpose, and they danced for the ancestors and the gods rather than for an audience composed of other ethnic or tribal groups. By contrast, in a Ujamaa village people danced on arbitrary occasions (presumably planned by a committee), there was a deliberate, reflected-upon political purpose for each dance, and the audience was made up of fellow villagers of different backgrounds. To reiterate, then, whatever its claims to the contrary, fundamentalism is *not* traditionalism.

WHAT DOES FUNDAMENTALISM LOOK LIKE ON A SMALL SCALE AND IN AN ENTIRE SOCIETY?

The fundamentalist project comes in two versions. In the first version, fundamentalists attempt to take over an entire society and impose their creed on it; in other words, they want to make the fundamentalist creed the taken-for-granted reality for everyone in that society. In the second version, fundamentalists abandon any attempt to impose a creed on everyone—the overall society is left to go to hell, as it were—but they try to establish the taken-for-grantedness of the fundamentalist creed within a much smaller community.

Let's call the first version of the fundamentalist project the *reconquista* model. The term *reconquista* was first used for the Christian "reconquest" of Spain from Muslim domination. Then, more relevant for the present argument, it was used again in the 1930s by Francisco Franco and his supporters in the Spanish Civil War of the 1930s. This time Spain was to be reconquered not from Islam, but from Communism, atheism, and all the other alleged deformations of the modern age. This, claimed Franco, would lead to a restoration of the *siglo de oro*, an imagined golden age during which the society had been "integrally Catholic" and authentically Spanish.

For the *reconquista* model to have a chance of success,

fundamentalists must have total or near-total control of all communications that could subvert the fundamentalist world-view. Put differently, the relativizing forces of modernity must be kept at bay. And this has a necessary institutional requirement—namely, the establishment and maintenance of a totalitarian state. It's important to understand that this is a much more radical political phenomenon than mere authoritarianism, as Hannah Arendt and other recent analysts have made clear. The authoritarian state doesn't brook political opposition, but it leaves people more or less alone as long as they go along with the regime. By contrast, the totalitarian state seeks to control every aspect of social life. It's not enough to eschew political opposition; one must enthusiastically participate in every activity set up by the regime.

The word "totalitarian" was coined, in a very approving tone, by Italy's Benito Mussolini. In one of his early speeches, he proclaimed the proposition that the Fascist regime was "totalitarian"—its basic principle was that there was to be "nothing against the state, nothing without the state, nothing outside the state." (The irony here is that Fascist Italy was authoritarian rather than totalitarian, but that's another story.) Mussolini's formulation is a pretty good description of the totalitarian state. In the twentieth century, there were two major cases in point—Nazi Germany (though Arendt argued that it became fully totalitarian only after the outbreak of World War II) and the Soviet Union and its various imitators outside Russia. Ideally, the totalitarian state provides institutions to control every phase of an individual's life from cradle to grave, and that control is reinforced by an ongoing barrage of propaganda and by agencies of state terror.

It can be said quite confidently that totalitarianism failed in the twentieth century. The Nazi case was destroyed by outside forces by means of war, of course. It would be counter-factual to speculate how Nazism would have evolved if Hitler had won the war or if there had been no war. But the Soviet case is more instructive. Soviet totalitarianism collapsed from within—there was no act of unconditional surrender on an American warship, no Allied military government set up in Moscow. The collapse of the Soviet Union certainly had multiple causes, such as the inherent failures of its socialist economy (including the failure to keep up with the United States in the arms race), the costs of empire (culminating in the campaign in Afghanistan), and the corruption and loss of nerve of the ruling elite. But the regime's failure to control outside communications also contributed, because it undermined the ideological monopoly of the regime. Modern means of communication make such control very difficult indeed, especially if a totalitarian regime wants to develop its economy. To be economically viable, it must deal with the outside world. And if the regime tries to limit communications with the outside to what's necessary to maintain economic relations, other communications still have a way of creeping in and creating cognitive dissonance with regard to the official creed. Eventually the regime opens up—economic *perestroika* leading to cultural *glasnost*—and then all the barriers against cognitive dissonance break down. It's fairly clear by now that the collapse of totalitarianism doesn't necessarily lead to democracy and pluralism. But it's likely to lead from a totalitarian state to an authoritarian one—which means that the fundamentalist project has failed in terms of embracing the society as a whole. Similar processes can now be observed in China (though the regime there is trying, so far successfully, to resist the more dangerous forms of *glasnost*).

As these examples show, the *reconquista* version of fundamentalism is exceedingly difficult to sustain, at least under modern conditions. Totalitarian regimes try to build walls against pluralizing and relativizing communications from the outside, but the powerful forces of the modern global economy batter at these walls until, sooner or later, there are breaches that allow massively subversive communications to pour in.

These difficulties don't mean that totalitarianism is impossible under modern conditions. It *is* possible—but only at an enormous cost to the society involved. That society has to cut itself off from the outside world not only culturally but economically, bringing about widespread misery. North Korea today is a prime example of this. It's also an example of a further requirement—indifference by the ruling elite to the miserable conditions under which most of its subjects must live. But even if total isolation is achieved—misery be damned—such a regime is likely to be unstable.

We turn now to the other version of fundamentalism—what we might call "subcultural" or "sectarian" fundamentalism. This could be described as a micro-totalitarianism: Just as in the macro version, there must be rigorous defenses against the cognitive contamination that outside contacts threaten to introduce. Informational isolation is quite difficult to achieve under modern conditions, as we've noted, but it's easier in a subgroup than in an entire society.

The sociology of religion has been interested in the phenomenon of sectarianism ever since the classical works on this topic by Ernst Troeltsch and Max Weber were written around the turn of the twentieth century. As we saw in the first chapter, a distinction has typically been made between the two sociological forms of religion, the church and the sect: The church is a broadly based institution, into which people are born; by contrast, the sect is a small enclave within society, which people join as a matter of decision. This typology is helpful in sorting out different religious phenomena, but it can also be applied to cognitive minorities whose worldview is not religious. Thus, for example, people who believe that the earth is regularly visited by extraterrestrial beings form quasi-sectarian groups that protect members from the disconcerting views of the majority. But every sect contains a built-in psychological contradiction: It seeks to maintain cognitive taken-for-grantedness, but at the same time it's constituted by virtue of individual decisions— and every decision is, by definition, *not* taken for granted and therefore potentially reversible.

Ideally (from the standpoint of a subculture's survival), a sect is physically isolated from the cognitive majority. This is best done in bucolic circumstances, removed from the temptations (cognitive as well as behavioral) of urban life. In American religious history, good examples of this are the Amish, the Shakers, and the Mormons (after their migration to Utah). In addition, there are secular analogues in various utopian movements, such as the Oneida Community. If a flight into rural isolation isn't feasible, for whatever reason, settlement in compact urban neighborhoods is helpful to the group's survival. The ultra-Orthodox Jewish neighborhoods in Brooklyn and in Jerusalem are cases in point. In either case—whether the milieu is rural or urban—the social situation is designed to make it very difficult for an individual to "escape."

While totalitarian states set up watchtowers and electric fences to prevent people from crossing the borders into enemy territory, subcultures set up mental equivalents of these defenses. It requires a huge effort by an individual to jump over these internal barriers, and even if that escape has been *physically* successful (after all, an ultra-Orthodox Jew can just walk outside Williamsburg, in the borough of Brooklyn, and take the subway into Manhattan, or take a bus to get away from Jerusalem's Mea Shearim quarter), "runaways" generally feel lifelong guilt at having betrayed their heritage and the people who embody it (parents, family in general, old friends, teachers).

If a subculture survives for longer than one generation, there will be an obvious difference between those who were born into it and those who joined it through a conversion experience. For the first group, necessarily, the subcultural definitions of reality have indeed acquired a degree of taken-for-grantedness. The sect will have acquired some "churchly" characteristics, so to speak. For the second group, the taken-for-grantedness must be laboriously constructed and vigorously maintained. For this reason, converts are typically more fervent than "natives." Put differently, the "native" has been socialized from childhood into the worldview of the subculture; the convert must be *re*socialized into this worldview.

This social-psychological change was described by Max Weber in his marvelous concept of the "routinization of charisma." The term "routinization" is a quite felicitous translation of the German *Veralltäglichung*. A literal translation would be "everydayization"—a terminological barbarism not even sociologists would want to be guilty of. Problems of translation apart, the process in question is clear: As time passes, the astounding character of the charismatic event weakens and everyday reality asserts itself. The astonishment gives way to routine and habit. The extraordinary becomes ordinary again. In terms of the sociology of religion, this can be described as a process whereby sects become churches. The same dynamic can be found in sectarian groups that propound secular rather than religious beliefs and values.

WHAT REQUIREMENTS DO FUNDAMENTALIST GROUPS TYPICALLY IMPOSE?

Any sect or subculture, whether religious or secular, has two broad requirements for "conversion" (no matter whether that conversion was sudden or gradual, voluntary or coerced). These requirements form the defense mechanisms mentioned above—the mental equivalents of the defended borders of totalitarianism. These same requirements apply to those who belong to a subculture by virtue of birth rather than individual choice, except that in their case the requirements are part of the taken-for-grantedness; the entire milieu ongoingly and silently prevents escape. Thus if one describes the requirements for conversion, one is at the same time describing the mechanisms by which "natives" are prevented from jumping over the wall.

The most basic requirement is the same in all sects, and

it echoes the requirements of a totalitarian state: *There must be no significant communication with outsiders.* As we saw earlier, the Apostle Paul understood this very well: He admonished early Christians not to be "yoked together with unbelievers." Anthropologists would translate this maxim into a prohibition of commensality and connubium—in other words, don't eat with unbelievers, and especially don't marry them! But human beings have a deep need for communication. Therefore, the subculture must satisfy this need through intense interaction within itself. Typically this includes providing the social situations where acceptable marriage partners can be found—that is, potential spouses who share the subcultural worldview.

The behavioral component of this requirement is simple: As we just saw, sects isolate their members—preferably in a rural setting, but sometimes in the middle of a city. There's a cognitive component to this requirement as well: Sects paint outsiders as ignorant of the "obvious" truth of the subcultural worldview. For the "native," the world has been divided from birth into a sharply dualistic scheme—those on the inside, who dwell in the light of truth, and those on the outside dwelling in the darkness of ignorance. (If that ignorance results from a *deliberate* rejection of the truth, the outsider is not just to be pitied but must also be condemned.) For the convert, coming from the pool of the ignorant, this entails a bifurcation of biography: The convert's life is reinterpreted in terms of a period BC and a period AD—that is, before and after the conversion. The preconversion period is defined pejoratively, of course. Converts frequently blame their parents for this period of darkness and very often break all previous family ties: They can't *afford* to remember.

A secular illustration of reinterpreting one's life in the wake of a conversion is provided by the Communist technique of brainwashing. This technique was first developed in the Soviet Union but was probably perfected by the Chinese Communists. It was employed by these regimes both in the training of cadres and in the treatment of prisoners (in "reeducation" camps, for example, and in prisons holding captured soldiers). In one version of the technique, individuals were given the assignment of writing down their own life stories. These texts were then "corrected" by the instructors and returned to the authors for revision. The authors were required to rewrite their narratives as many times as it took to "get it right"—that is, until an autobiography had been reconstructed in accordance with Communist ideology (for example, until it had been purged of "remnants of bourgeois consciousness"). In cadre training, presumably individuals underwent this exercise voluntarily—either because they sincerely believed in it, or because they were opportunists. Prisoners, on the other hand, had to be coerced into the exercise. Typically this involved a period of physical maltreatment and degradation: The old identity was broken down so that a new one could be constructed. In various forms, similar albeit more benign techniques have been used in other settings—in the training of novices in monastic orders, in basic training in the military (especially in elite units, such as the Marines), or in psychoanalysis (at least of the classical Freudian type, which can be described as a prolonged rewriting of the patient's biography—until, finally, he or she "gets it right").

The second requirement for conversion into a sect builds on the first, and it likewise mirrors totalitarianism: *There must be no doubt.* Fundamentalists, in particular, can't tolerate doubt; they seek to prevent it at all costs. As with the first requirement, this one has both a cognitive and a behavioral component. Cognitively, the suppression of doubt is addressed primarily in socialization. "Native" inmates of the community have been drilled in the sect's particular ideology from childhood; converts, newly socialized, must be watched especially carefully (and must watch themselves), in order not to "relapse" into earlier habits. If questioning should occur despite the preventive measures, therapeutic practices designed to deal with doubts can be applied. In a religious context, these might be called "cure of souls" or just plain "pastoral care."

For "natives," socialization begins in childhood, when a worldview is inculcated by so-called significant others—that is, persons of great emotional importance for the child. Most likely these are first and foremost the parents, but others can also fulfill this role—older siblings, other revered relatives or friends, teachers, or clergy. Converts, of course, lack correct primary socialization; they must be *re*socialized. Significant others are just as important to the convert as to the "native," however, as are what we might call conversion personnel— spiritual directors, party functionaries, drill sergeants, or psychoanalysts. The convert typically forms intense personal relationships with these people. In psychoanalysis, this process is called "transference," to use a term coined by Freud (though we're clearly deviating from the meaning intended by him): We can say that there are significant others who assist the convert to "transfer" from one worldview to another, and who help the convert to stay in the latter. Not surprisingly, the individual is infantilized in relation to these authority figures; in a psychological sense, it's a reversion to childhood.

Doubt-resisting behaviors have cognitive correlates. Conversion personnel are aware of this. Thus spiritual directors, when confronted with individuals afflicted with doubts, will first of all recommend certain behaviors. The formula here is simple: One need not believe in order to pray; one prays *in order to believe*. In addition, there are cognitive mechanisms to assist in the process of doubt-containment. Generically, these fall into the categories of nihilation and apologetics.

Nihilation is generally the cruder mechanism: Doubts are dispelled by being given a negative status. In religious cases, they are subsumed under the heading of sin; thus a lack of faith is deemed sinful, a rebellion against God. We've already mentioned the wonderful Communist notion of "remnants of bourgeois consciousness," and psychoanalysts speak of "resistance"—both labels of negativity. Such nihilation exercises make it unnecessary to deal with dissonant definitions of reality on their own terms; those definitions can be dismissed as unworthy of serious consideration.

Apologetics (to use the Christian theological term), on the other hand, may be crude or highly sophisticated. In either case, they are a body of arguments intended to defend the validity of the sectarian worldview. In the most ample form they supply a full-blown theory—this or that theological system, or Marxism, or Freudian psychology: Doubts are liquidated by immersion in a comprehensive theory that both explains and denies them.

WHAT IS THE ULTIMATE COST
OF FUNDAMENTALISM?

Every worldview *locates* the individual. Put differently, every worldview provides an identity. Fundamentalism does this in both its *reconquista* and its subcultural versions. This identity is intended to be taken for granted, to be endowed with self-evident validity. The individual now *is*, or (in the case of the convert) *becomes*, what he or she is supposed to be. In its political manifestation, Erich Fromm aptly called this the "escape from freedom," as we noted in an earlier chapter. If one values freedom, and a society in which freedom has become institutionalized by liberal democracy and the constitutional state, then fundamentalism must obviously be seen as a serious threat. Fundamentalism, religious or secular, is always an enemy of freedom.

We argued earlier that relativism undermines the "collective conscience" and thus the solidarity (Durkheim) of a society. But so does fundamentalism. In its *reconquista* version it does indeed seek to create solidarity based on a coerced uniformity of beliefs and values. But the totalitarian regime that must be established in order to maintain such conditions has enormous economic and social costs. The subcultural version of fundamentalism seems to have fewer costs. At least initially, while the subculture is both small and rare, the costs are borne only by its members. But if such subcultures multiply, they undermine the cohesion of the society, which becomes "balkanized." Then the costs are borne by everybody. The final outcome may be all-out civil strife, between radicalized subcultures and the majority society, and/or between/among the several subcultures themselves.

If the danger of relativism to a stable society is an excess of doubt, the danger of fundamentalism is a deficit of doubt. Both extreme uncertainty and extreme certainty are dangerous, though not equally so. As to the dangers of certainty, it's interesting to consider the imposing figure of Oliver Wendell Holmes. A member of the Boston cultural and social elite of the nineteenth century, he served, as a young man, in the Union Army during the American Civil War. He was horrified by the atrocities committed by both sides. He returned from the war with the conviction that *any* certainty is nefarious and potentially dehumanizing. He felt that, conversely, skepticism (doubt as habit, if you will) is essential to a humanly decent society. This conviction influenced his actions as a justice of the U.S. Supreme Court.

We agree with him. It follows that one ought to establish a middle position, equidistant from relativism and fundamentalism. The religious and moral aspects of such a position are similar but not the same. These issues will be taken up in the next few chapters.

5

CERTAINTY
AND DOUBT

The twentieth-century Austrian novelist Robert Musil once remarked, with the irony characteristic of his writing, "The voice of truth has a suspicious undertone." That statement calls to mind Pascal's previously mentioned remark that the truth on one side of the Pyrenees can be an error on the other side. Truth, in other words, is less certain or absolute than the "true believer" wants. To rephrase that idea philosophically, truth stands open to falsification. It's related to time and space—though a person who adheres to one or the other metaphysically grounded belief or faith doesn't see it that way. Through most of history such grounding was supplied by religion, and it's in the sphere of religion that the interplay between certainty and doubt has been acted out most dramatically (as will be discussed below). Today, though, there are many "true believers" without any religious affiliation. In other words, there's a truly ecumenical community of fanatics of every persuasion, religious *and* secular.

AREN'T THERE SOME TRUTHS
THAT ARE ABSOLUTE?

The fact that truth stands open to falsification doesn't mean that there's no "doubtless" truth at all. There are, to begin with, the mathematical rules of thumb that no one in his right mind would subject to doubt and falsification. In all ages and in all times, it's obvious that two added to three is five, and two multiplied by three is six. This is a common-sensical truth that any sane person ought to take for granted. Yet the truth of mathematics, like Pythagoras's proposition, must be taught and learned. It's not innate and naturally self-evident. In fact, it's quite difficult to give a convincing verbal answer to the question of a child who asks why four added to four is eight and not nine or seven. The only thing to do is to take the hands of the child and start adding the fingers. Incidentally, for children of, say, five or six years old, who have already been taught the beginnings of arithmetic, it's a wonderful discovery to find ten fingers and ten toes that can be counted, added, deducted, and multiplied. In fact, the fingers of a child function as a primitive abacus.

Truth and madness are sometimes antagonistic but strangely joined twins. The social psychologist Milton Rokeach, in the early 1960s, studied three mental patients in three different institutions, each of whom believed himself to be Jesus Christ. Rokeach thought they might be cured from this delusion by bringing them together in one institution. They then would be confronted with serious cognitive dissonance, since the existence of three Christs is an obvious impossibility—apart from the equally obvious fact that Jesus is since roughly two millennia no longer physically among us. With the help of their psychiatrists, who remarkably followed Rokeach's hypothesis of the healing potential of the to-be-inflicted cognitive dissonance, the three were brought together in one asylum in Ypsilanti, Michigan. Rokeach recorded the often heated debates of the three men and published them in his book *The Three Christs of Ypsilanti* (1964). At one point Rokeach thought that the most intelligent of the three began to heal indeed from his delusion. The man said that in his view the other two men had to be raving mad, since they believed themselves to be Jesus Christ. This, he continued, is really absurd because naturally only one person can be Jesus Christ. Sure enough, not they but he was the Christian Messiah.

Rokeach also related another case of a conflicting plural identity. Two patients, an older and a younger woman, both believed herself to be Mary, the mother of Jesus. They were quarrelling all the time, until suddenly the older woman found a solution. She asked the doctor who the mother of Mary was. Upon some reflection he said that, if he was not mistaken, the mother of Mary was called Anna. The older lady then happily announced that she was Anna, warmly embraced her younger fellow patient, and addressed her from then on as her daughter Mary. Cognitive dissonance was thus ingeniously dissolved.

Close to the objective and indubitable rules of arithmetic and mathematics are those of formal logic. In fact, according to most philosophers, mathematics and logic are intrinsically cognate. This isn't the place to enter into the highly specialized and difficult field of logic. It suffices to say that there are basic propositions in logic, such as the syllogism, that are generally held to be beyond doubt—that is, to be absolutely true. "Human beings are mortal; Socrates is a human being; therefore, Socrates is mortal." This is an example of the most basic syllogism, and it contains an indubitable truth. In fact, substantive terms such as "human being," "mortality," and "Socrates" can be substituted by immaterial symbols, as has happened in logic ever since Aristotle: M is P; S is M; therefore, S is P. In fact, formal logic *prefers* such symbols, since it's not concerned with philosophical or theological ruminations—especially not if they're about human mortality and immortality, as in the case of the above-mentioned syllogism.

But none of this addresses the truth and certainty we're longing for in everyday life. Life isn't a sum of formal syllogisms but an often painful succession of choices and decisions pertaining to alternatives that aren't at all "rational," nor are the choices and decisions "logical." Formal logic tries to eliminate the undertones of Musil's "voice of truth," but in life these undertones are very hard to eliminate. Truth is perpetually overshadowed by doubt and insecurity. Only the "true believer" who has settled down in one or the other religious or philosophical "–ism" will shout down the voices of doubt—voices that, as we saw before, are multiple in the ongoing pluralizing process of modernization.

Yet there is in fact a very fundamental *non*-religious and *non*-philosophical certainty in our lives, one that offers what Arnold Gehlen aptly called "a benign certainty." We discussed Gehlen's work in the first chapter but must briefly return to it. His "benign certainty" is the largely taken-for-granted certainty of institutions, which are transmitted from generation to generation, establishing what's called "tradition." Marriage, family, church, temple, mosque, school, university, voluntary association, etc.—all these are not just functioning organizations. They're also meaningful institutions that carry the values and norms that give direction and certainty to our daily actions and interactions.

When, for example, one migrates to a foreign country, one ought to learn its language, its manners, its religious and secular ceremonies, its ways of acting, thinking, and feeling—in short, its institutions. In that way, one appropriates the meanings, values, and norms of the people in the new social habitat. Such appropriation is necessary if one is to communicate and interact with one's new neighbors. It may take a while, but eventually one will experience the "benign certainty" of an institutionally grounded taken-for-grantedness. It's a sense of feeling at home, although the old world one emigrated from lingers on in memories and emotions. In fact, the sense of living in between two different worlds is often a twilight zone of multiple doubts and uncertainties that lasts until one's death. Usually, though, it fades away in the second or third generation. Migration isn't a new phenomenon, but it's reached unprecedented dimensions in the modern era. Thus the world today contains millions of people who straddle two, and often more than two, cultures.

As we've argued before, the modern process of pluralization has been a deinstitutionalizing and existentially destabilizing force. It has enlarged our freedom of choice and thus in a sense our autonomy and self-reliance. Yet, as a visit to any modern supermarket demonstrates, we're confronted also by the *Qual der Wahl* ("agony of choice") that we mentioned in the second chapter. In fact, the supermarket can be taken as a metaphor of a fully pluralized society. This pluralization has led to two opposite reactions. There is, on the one hand, a radical return to premodern certainties, such as religious fundamentalism and scientific rationalism, and, on the other hand, an often equally radical celebration of allegedly postmodern contingencies, which are propagated as a relativism in which (morally) "anything goes." In the first case, the agony of choice is mitigated by the introduction of a theological or philosophical canon of truth. In the second case, it's simply turned into an alleged advantage, since relativists believe that choice is the ultimate guarantee of freedom and autonomy.

Neither position is plagued by doubt; they have that in common. Theirs is a certainty that—allegedly—is indubitable. In fact, both positions are held by "true believers" who find their certainties in religion, in science, or in postmodernist relativity. The latter in particular often claim to celebrate doubt, but they in fact absolutize doubt into a radical relativism or cynicism that heralds the *end* of doubt. Actually, relativists and cynics too are "true believers." What is it, then, that makes a "true believer"?

HOW DO "TRUE BELIEVERS" HANDLE DOUBT?

In 1951 the American longshoreman and cracker-barrel philosopher Eric Hoffer published a little book entitled *The True Believer*, in which he presented a profound picture of this type of human being. Mass movements, he argued—like religious, social-revolutionary, and nationalist movements—propagate very different ideologies yet share one characteristic that gives them a family likeness: They generate and are carried by people who in extreme cases are even prepared to die for the cause, who call for conformist action, and who cultivate and are driven by fanaticism, hatred, and intolerance. Quite apart from the doctrines they preach and the programs they project, these mass movements share the same type of mind—that is, the fanatic mind of the true believer. Hoffer saw this type of mind in Christian and Muslim radicalism (nowadays we speak of Protestant fundamentalism and Islamism), in Communism, in Nazism, and in various forms of nationalism. The following observation, made by Hoffer in 1951, is still valid: "[For] though ours is a godless age, it is the very opposite of irreligious. The true believer is everywhere on the march, and both by converting and antagonizing he is shaping the world in his own image." As in Hoffer's day, there are many other "–isms" that have been created and propagated by true believers, such as Enlightenment modernism, antirational romanticism, and equally antirational postmodernism.

Most of these "–isms" can be called "gods"—that is, objects of devotion and worship—though a Hebrew prophet would define them as "*false* gods." Often they are "gods that have failed," to paraphrase the title of a volume of essays by six European intellectuals, who in the period between 1917 and 1939 believed in the blessings of Communism but lost their belief after having become acquainted with Stalin's terrorist brand of it. Such secular gods fail in particular when their prophecies fail, as when the proletarian revolution, the predicted end of the world, or the prophesied return of the messianic figure doesn't happen. In early Christianity, there were high expectations of the imminent return of Jesus Christ to establish the Kingdom of God on earth. It has been suggested that this frustrated eschatology stimulated the missionary activities of the Apostle Paul and the establishment of the Christian church as a formal organization. The pontiff of Rome was seen not just as the chief executive officer of the Roman Catholic Church, but also as Christ's deputy on earth until Christ's return.

In his classic study *When Prophecy Fails* (1956), Leon Festinger argued that people who are deeply committed to a belief and its related courses of action won't lose this belief when events falsify its assertions, as when a prophesied event fails to occur. On the contrary, they will experience a deepened conviction, and start to proselytize in order to receive further confirmation of their belief. The more people embrace their belief, the truer this belief must be—or so their thinking goes. However, Festinger adds, in most cases there comes a moment when the disconfirming evidence has mounted to a degree that obstinate doubt creeps in. That doubt, as it grows, eventually causes the belief to be rejected—unless, that is, the believers succeed in a solid institutionalization, as has been the case with Christianity. The dissolution of apocalyptic movements is more likely when a precise date for the end of the world has been given (and has passed). Sooner or later, after this date has expired without any apocalyptic disaster having occurred, such a movement generally collapses (though one must not overlook the capacity of human beings to deny disconfirming evidence).

Religious and secular fundamentalists and their opponents have engaged in bitter controversies throughout recorded history. Though these groups are diverse, they typically share three main characteristics: First, they have great difficulty listening to opposing opinions and ideas. Second, they claim to possess an irrefutable truth (whether religious or secular). Third, they claim that their truth is the *only* truth; in other words, they declare a monopoly on truth. The opposing positions of "creationists" and "evolutionists" present a telling example. If such quarrels were to remain restricted to the inner spaces of institutions such as the church, the mosque, the temple, the synagogue, or the university, they would be relatively harmless. Yet true believers fight in public places—in the political arena in particular—where they can cause considerable harm.

Because of their conviction that they have a monopoly on truth, "true believers" repress even a hint of doubt. They ridicule or even persecute those who represent liberal moderation. Religious fanaticism was what caused Voltaire to exclaim *"Écrasez l'infâme"* ("Destroy the infamy")—the infamy being the church and perhaps Christianity in general. But the Enlightenment produced its own murderous fanaticism. Not long after the goddess of reason was enthroned by the French Revolution (in the Church of the Madeleine, no less), the Terror was unleashed, easily surpassing the cruelties of the *ancien régime* that had outraged Voltaire.

The religiously motivated suppression of doubt can be exemplified by one of the many historical examples presented by centuries of European religious conflicts. The sixteenth-century French reformer John Calvin, a true believer if there ever was one, established a Protestant theocracy in the city of Geneva. He never aspired to a political position but remained a church minister throughout his life. Yet, as a kind of *ayatollah*, he tried to get a firm grip on the political arena of his city. At first he wasn't successful: The mayor and the city council refused to surrender to the doctrines of Calvin and his equally fanatic colleague Guillaume Farel, and banished them from Geneva. But two years later, after the factional relations within the council had altered, the two reformers were asked to return to Geneva. Calvin then issued his *Ordonnances ecclésiastiques*, which was accepted by the city council. In these strict ecclesiastical regulations, Calvin introduced a radically presbyterian church council—one consisting of theologically trained ministers and lay elders—rather than the episcopalian, or bishop-governed council, that had been customary. Remarkably, these laymen were also to be governmental officials. The church had to be autonomous, Calvin ordered, yet the city-state was to be subject to the church, particularly in the broad field of morality. Calvin's theocracy was plagued by strife and conflict, yet he managed to hold a firm doctrinal and moral grip on the citizens of Geneva.

Naturally, Calvin was opposed fiercely by some theologians. Among the first was Cardinal Jacopo Sadoleto, secretary to Pope Leo X and bishop of Carpentras, in southern France. In a letter addressed to the city council and the citizens of Geneva, he tried to persuade these Protestant "heretics," Calvin and Farel, to return to the mother church. Using as his trump card the issue of salvation, Sadoleto posed the following question. What will happen to our soul—the essence of our identity—after we die? Damnation or salvation? The Holy Roman Church, having existed for more than fourteen centuries (Sadoleto continued), offers salvation through the eucharist, the confession of sins and their absolution, the prayers of the saints to God on our behalf, and our prayers to God for the dead. Naturally we're in need of the grace of God, but good works are equally necessary for our salvation. Sadoleto's tone was conciliatory, yet at times he erupted in fanatic anger: "For well knew I, that such innovators on things ancient and well established, such disturbances, such dissensions, were not only pestiferous to the souls of men (which, is the greatest of all evils), but pernicious also to private and public affairs." But he ended his letter on a conciliatory note: "I will not, indeed, pray against them that the Lord would destroy all deceitful lips and high-sounding tongues; nor, likewise, that He would add iniquity to their iniquity, but that He would convert them, and bring them to a right mind, I will earnestly entreat the Lord, my God, as I now do."

Although the letter wasn't addressed to him, Calvin responded to it within five months with a lengthy, well-written exposition. He started by saying that his activities in Geneva were not intended to promote his private interests, as Sadoleto had suggested. Everything he was doing, he said, was being done in service to Jesus Christ and in obedience not to the church but to the Bible, "a cause with which the Lord has entrusted me." In fact, he added, "if I wished to consult my own interests, I would never have left your party. I certainly know not a few of my own age who have crept up to some eminence—among them some whom I might have equaled, and others outstripped." And with all due respect, he continued, it was somewhat suspicious that a person who had never been in Geneva and had never shown any interest in the Genevese "now suddenly professes for them so great an affection, though no previous sign of it existed." What the bishop intended was, Calvin saw, "to recover the Genevese to the power of the Roman Pontiff." And as to salvation, it was in Calvin's view not sound theology to confine one's thoughts and fears so much to oneself—that is, to one's soul; instead, one should primarily demonstrate the glory of God. Yes, works are important, though not in order to reach heavenly life but to honor the glory of God. Furthermore, we should not confuse this glory with that of the pontiff of Rome and his subordinates. Salvation can be reached only by faith and the mercy of God: "We show that the only haven of safety is in the mercy of God, as manifested in Christ, in whom every part of our salvation is complete. As all mankind are, in the sight of God, lost sinners, we hold that Christ is their only righteousness, since, by His obedience, He has wiped off our transgressions."

At one point in his epistolary response, Calvin bragged of his "conscientious rectitude, heartfelt sincerity, and candor of speech." In comparison to Sadoleto, he even claimed to have been "considerably more successful in maintaining gentleness and moderation." This was written in August 1539. He certainly had lost this "gentleness" when he published his *Catechismus Genevensis* six years later, in which he unfolded the theocratic, disciplinarian regime to which the citizens of his city were to be subjected. By that point, Calvinism had become an established ideology. All criticism was fanatically suppressed, but the critics couldn't be silenced.

The most vociferous reaction came from Sebastian Castellio, who initially was a close friend of Calvin's, but who moved away from the latter's increasing fanaticism, criticizing especially Calvin's doctrine of predestination. Castellio eventually fled from Geneva to Basel, where he intensified his anti-Calvinistic emphasis upon toleration and freedom of conscience. In 1553, Michael Servetus, a somewhat confused lay theologian who rejected the doctrine of the Trinity, was publicly burned at the stake—a very painful death to which Calvin objected, unsuccessfully proposing the rope as an alternative ("gentleness" indeed!). In response to this atrocity, Castellio published two treatises in which he passionately rejected the persecution and execution of heretics. He distanced himself most radically from theological fanaticism by his treatise on "The Art of Doubt, Faith, Ignorance and Knowledge" (1563). In it, Castellio tried to answer the multifaceted question, Which Christian doctrines ought to be subjected to doubt, which ought to be believed, which does one not have to know, and which ought one to know? Though the treatise was wide-ranging, doubt was the issue Castellio was most interested in, in opposition to fanatics like Calvin.

There are, he argued, many passages in the Old and the New Testaments that are hard to believe and stand open to doubt. There are, for instance, many contradictions that open the gates of doubt. Yet—and here Castellio engaged in a remarkably early form of modern hermeneutics—we should deal with our doubt by focusing on the main drift, the spirit of the words in the context of their coherence. Doubt and uncertainty thus pave the road to knowledge and indubitable truth. Now, there's a category of people, he continued, who insist that one should not burden oneself with uncertainty, who assent uncritically to everything scripturally recorded, and who condemn without hesitation anyone who holds a different view. Moreover, these people not only never doubt, but they can't allow doubt to arise in the mind of someone else. If someone continues to doubt, fervent believers don't hesitate to call him a skeptic, as if someone who doubts *anything* would claim that *nothing* can be known or experienced with certainty. Castellio paraphrased Ecclesiastes 3:2, saying, "There is a time of doubt and there is a time of faith; there is a time of knowledge and there is a time of ignorance."

The most interesting part of Castellio's theory is his juxtaposition of ignorance and knowledge on the one hand, and doubt and faith on the other. He saw ignorance as an unavoidable preparatory stage for knowledge; likewise, he saw doubt as a preparation for faith. Moreover, and this is a remarkable dialectical step, he saw ignorance and doubt not as the totally different opposites of knowledge and faith, but as intrinsic counterparts. This, of course, runs counter to the worldview of true believers of every day—not only in the world of religion but in that of rationalism as well. As to the latter, Castellio seemed to foresee the rise of scientific rationalism as an inherent component of the process of modernization.

And he certainly foresaw correctly: With the rise of the sciences in the Western world, we have witnessed the birth of what has aptly been called "scientism"—the often quite fanatic belief in the omnipotence of the (mainly natural) sciences and their technological applications. It is a form of rationalism that fanatically fights all forms of alleged ignorance—but that of religion in particular. While religious faith is defined as irrational ignorance, the rational sciences (including the social sciences, modeled after the natural sciences) are elevated to well-nigh metaphysical heights.

Auguste Comte, whom we introduced in a previous chapter, was an early representative of this rationalist worldview, which he labeled "positivism." Psychological and sociological behaviorism exemplified this ideology in the former century. It's still manifest in the natural sciences, albeit in the new cloak of geneticism. The "god" of this rationalism nowadays is "the selfish gene," which is a late-modern specimen of predestination. Like its Calvinist predecessor, it destroys the idea of freedom of the will and morally good works.

Doubt as an intrinsic part of faith is, of course, always rejected by true believers—those whom we nowadays label "fundamentalists." To the fundamentalist true believer, faith is not, as Paul Tillich defined it, "believing the unbelievable," but the reliance on God's or Allah's indubitable revelations, as recorded in sacred books, contained in sacred traditions, and experienced in sacred ceremonies. There's an interesting coalescence of scientism and religious fundamentalism in the ongoing debate between evolutionism and creationism.

WHAT, THEN, IS DOUBT?

Doubt is a rather complex phenomenon—multifaceted and pluriform. First, both superficial and profound doubt exist. When at the end of a sumptuous dinner a delicious desert is offered, it's hard for someone with a notorious sweet tooth to decide between yes or no. Economists might expect a *rational* choice, but that's unlikely. One knows it's better for one's health—and therefore rational—to say no, but it's common knowledge that temptations are hard to fight rationally. Oscar Wilde knew the solution for this type of superficial doubt: "The only way to get rid of a temptation is to yield to it." A profounder and more tormenting doubt might befall a bride or groom—allegedly not a rare case— shortly before the nuptial ceremony and festivities. "Should I really commit myself to this marital relationship, to this person—'for better or for worse, until death do us part'?" Or take another example: Recently several general practitio- ners in the Netherlands expressed their tormenting doubts about the practice of legalized euthanasia in that country. In the Netherlands, euthanasia remains subject to criminal law, but under some very strict conditions (and in cases of a terminal disease with terrible suffering) physicians can be exempted from prosecution. Yet many physicians struggle with serious and profound doubts when patients (and often patients' close family members) beg them to put an end to their suffering.

There's the kind of doubt that one fights or tries to avoid, as in the case of religious faith or some particular political- ideological belief. In that arena, true believers experience doubt as a possible road to apostasy. But there's also doubt that one wants to revel and immerse oneself in, as in the case of cynicism. A cynic can be defined as a person who sublimates doubt into a mode of thinking and a style of life. According to the cynic, everything and everybody must be constantly subjected to doubt, since nothing and nobody can be held to be true and trustworthy. Most cynics, by the way, believe that this rule doesn't apply to themselves— which *doesn't* mean, of course, that cynics are trustworthy and truthful.

While cynical doubt is usually tight-lipped if not grim, there's a kind of doubt that's playful and is expressed in jests and banter. In that case, doubt is generally couched in irony—in other words, one says one thing but means something else (with the second meaning containing some sort of critique and thereby shedding doubt on taken-for-granted beliefs). A famous example is Erasmus's *In Praise of Folly*, which is more than an exercise in poking fun at medieval philosophy and theology; it's an ironic critique of medieval intellectuals. In accordance with late medieval fools and folly, Erasmus tried to say in this treatise that Wisdom sees Folly if it looks at itself in a mirror. But it's also the other way around, because when Folly stands in front of a mirror, Wisdom is reflected. As joyous and funny as Erasmus's *In Praise of Folly* is, it leaves the reader feeling uncomfortable. Erasmus repeats the Apostle Paul's argument about the wisdom of the world being folly in the eyes of God. Despite that link to religion, however, the disquieting fact of Erasmus's approach is the absence of any metaphysical reality. The rational boundaries between wisdom and folly are blurred. They evaporate, dissolving in a kind of cognitive and existential fog. A very profound and even uneasy doubt arises from this fog.

In short, one can doubt big and important, or small and unimportant, things. One can harbor doubts about oneself, the world at large, or God. What these cases have in common is that they question whether something or someone is reliable, trustworthy, and meaningful—that is, whether something or someone is "true." Doubt and truth, in other words, are about relationships. In the next chapter, we will discuss this matter in more detail.

IS DOUBT AN
ALL-OR-NOTHING PROPOSITION?

When we're confronted by choices—and, as we saw in an earlier chapter, in modernity we're *constantly* confronted by *many* choices—doubt presents itself prominently. While choices can be as superficial as which shirt to buy when a consumer strolls through the shopping mall, or as serious as whether to administer a lethal dose of medication when a physician is confronted by the request of a terminally ill patient to put an end to his suffering, these extremes are *border* cases. Doubt is most common and most prominent as a *middle ground* between religious belief and unbelief on the one hand, and knowledge and ignorance on the other. These two opposites are, in fact, interrelated, as we just saw: Knowledge can foster unbelief, and ignorance can foster belief or faith. As to the latter, a medieval theologian introduced the notion of the *docta ignorantia*, the learned ignorance, as a method to deepen one's mystical sense of the divine. On the other hand, if one analyzes the sacred texts of religion scientifically—that is, historically and comparatively—one's faith may easily slide off in the direction of unbelief. The middle ground of all this is doubt—a basic uncertainty that isn't prepared to let itself be crushed by belief or unbelief, knowledge or ignorance.

Precisely because it occupies this middle ground, genuine doubt can never end up in the many "–isms" that people have invented and propagated. Doubt can't be relativistic, since relativism, like all "–isms," stifles doubt. Renaissance scholar Michel de Montaigne, in his famous essays, designed a pragmatic philosophy of everyday life, averse to metaphysics and religion. He struggled with the following paradox: He constantly stressed the relativity of human ideas, ambitions, projects, and activities, yet he refused to settle down in an easy relativism. In one of his essays, he remarked wryly that if one says, "I doubt," one should realize that one apparently *knows* that one doubts—and that is, of course, no longer doubt! (This brings to mind the Sophist logic about the Cretan who claims that all Cretans are liars, which claim then must be a lie.) However, Montaigne overlooked the possibility of doubting one's doubt! A Flemish poet once formulated the dilemma thus: "From the start the human condition seems to be determined as doubt about doubt." Notice the doubtful way in which doubt is defined here— "*seems to be* determined"—and then the quintessential notion that the human condition consists of doubt that doubts itself. That does, of course, open a door to knowledge and belief, but it's a stammering kind of knowledge and belief, not the knowledge and belief of the true believer. It *faces*, as it were, knowledge and belief, but it knows ignorance and unbelief at its back. Needless to add, this position is far removed from the cynicism and relativism with which it is often erroneously identified.

Actually, this kind of doubt is typical of a truly agnostic position. (The term "agnosticism" should be avoided, since agnostic doubt is alien to any "–ism.") The agnostic isn't an atheist. The latter, very much an adherent of an often fanatic "–ism," is a self-defined unbeliever who sets out to fight and attack any kind of religious belief and institution: *Écrasez l'infâme!* Politically, atheists defend a strict separation of church (mosque, temple, synagogue) and state, but many of them would prefer to eliminate by force all traces of personal and institutional religion. Stalin tried it, but it couldn't be done successfully, because his surrogate religion (Stalinism) lacked any sense of security—it could be enforced only by terror and mass murder. Indeed, it was a "god" that failed. Most atheists settle for the compromise of a separation of church and state, and restrict themselves to fighting their believing opponents in lecture halls, magazines, and newspapers. Since the discovery of DNA, they often elevate genetic structures and processes to metaphysical heights, as in the case of the previously mentioned "selfish gene." Darwin is to many atheists a kind of semi-religious prophet, the founder of a doctrine called Darwinism. In a recent debate about Danish cartoons that ridiculed the prophet Muhammad, a self-professed Darwinist atheist exclaimed emotionally, "Why are we not allowed to poke fun at the Muslim prophet, if Muslims and other creationist believers publish cartoons which depict my prophet Darwin as a monkey?" The funny thing was that he didn't mean this as a joke. He was the prototype of the atheistic true believer.

The agnostic position is by definition a weak one. The agnostic doesn't radically reject what's believed in religion, as the atheist does. Maybe he even would like to believe as the believer does, but the knowledge he's gathered by study and experience restrains him. Asked if he believes in or hopes for a life after death, the agnostic won't answer as the atheist does: "No, of course not; my death is the absolute end of me." The agnostic will murmur, "Well, I may be in for a surprise." Doubt is the hallmark of the agnostic. The believer might immediately answer that he too is confronted by doubt all the time, adding that that's why it's faith and not knowledge that he or she adheres to. The difference is that the believer is plagued by doubt and searches all the time to be delivered from it, whereas doubt is endemic to the agnostic. If not a fanatic true believer like Calvin, the believer lives with and in faith that is troubled by doubt. If not a fanatic atheist like the quoted Darwinist, the agnostic lives with and in doubt that is troubled by faith. It's a thin line, but an essential divide.

There's a causal relationship between rational critique and doubt. When Descartes, whose philosophy we touched on in chapter 3, exclaimed *"De omnibus dubitandum est"*—"Everything ought to be subjected to doubt"—he meant this as a methodological and epistemological device by which to arrive at truth in a rational manner. In other words, one ought to bracket the religious doctrines and metaphysical theories that have been piled up in the history of human ideas. Or, in contemporary parlance, one ought to clean the hard disc of one's mind by deleting the information passed on by former generations. Descartes intended to liberate us from what Montaigne called "the tyranny of our beliefs," and to foster what Montaigne described as "the liberty of our judgments." Montaigne and Descartes stood in the tradition of Socrates, who in his disputes with students asked questions in order to destroy preconceived ideas and fashionable (but often false) beliefs. He asked questions and claimed that he didn't know the answers. His aim was not to teach beliefs but to cleanse the mind of false beliefs, preconceived ideas, and prejudices. In other words, Socrates taught his students to subject their convictions systematically to fundamental doubt. This "logic" of doubt was picked up again by Francis Bacon, who designed a philosophy—or better, a methodology—that would serve the practical, scientific interpretation of "the book of nature" based on precise and detailed observations of reality. An absolute precondition was that we learn to discard the *idola*—that is, the fallacies we carry about in our minds, preventing us from acquiring a sound knowledge of reality. One group of *idola* he called the "fallacies of the market," which emerge in society where people converse with each other and engage in transactions of all sorts. Here, back four hundred years or so, is a foreshadowing of the sociology of knowledge. Another group, which he labeled "fallacies of the theatre," consist of the transmitted notions of ancient and medieval philosophers that have attached themselves as clichés to our minds. All these *idola* are to be subjected to systematic doubt.

It stands to reason that the Socratic-Cartesian-Baconian method opens the gates to a fundamental type of cognitive doubt. This doubt is in fact a line of rational-critical thinking that in the past century was picked up again by Karl Popper, when he introduced the notion that the hallmark of scientific research is not verification but, on the contrary, falsification. Theologians and metaphysical philosophers try to demonstrate the truth of their theories—verification—whereas critical-rational scientists open the results of their research to falsification: "Show me where my hypotheses and theories are wrong!" It's only in this way that our knowledge about the world can progress—step by step. It's an incremental kind of evolution of knowledge that is, however, constantly carried by doubt.

In his unfinished essay on the Cartesian slogan *De omnibus dubitandum est*, Søren Kierkegaard claimed that doubt is negative, since it always reflects critically on existing theories and ideas. "Doubt," Kierkegaard argued, "is to refuse one's approval. The funny thing is that I refuse my approval each time something happens." Doubt thus is in essence a reaction, and therefore not fit to function as the start of philosophy, as Descartes and others wanted us to believe. Rather, an attitude of wonderment about the world around us should be seen as the source of proper philosophical thinking. Kierkegaard found this wonderment, contrary to doubt, a positive attitude that's not reflexive but proactive. We might add that wonderment and curiosity also stand at the cradle of scientific research. However, it's questionable to oppose negative doubt and positive wonderment as Kierkegaard does, because wonderment doesn't occur in a neutral environment but is always surrounded by Baconian *idola*, which ought to be bracketed or wiped out in order that we become able to wonder about the world around us. The small child can approach the world in pure wonderment, unencumbered by any *idola*. But the adult is thoroughly socialized and enculturated in a reality that can't be naively taken for granted anymore. That's certainly the case in a fully modernized and therefore, as we saw before, pluralized society. In such a society, doubt and wonderment are, as it were, twins.

WHAT DISTINGUISHES SINCERE DOUBT FROM MERE CYNICISM?

Relativism, as discussed in chapter 3, and the cynicism we discussed above excel in forms of doubt that are logically inconsistent and morally reprehensible. We will discuss the moral dimensions of doubt in the next chapter. At this point, we must briefly reflect on the inconsistent nature of cynicism and relativism, and juxtapose that inconsistency against the consistent and sincere type of doubt. Relativism and cynicism subject everything and everyone to doubt, yet as "–isms" they generally harbor true believers, who—as true believers—don't apply this doubt to themselves! Kierkegaard pointed ironically at the inconsistency of the relativist. That someone decides to doubt is understandable, but it's *not* understandable that someone propagates doubt to someone else as the correct thing to do. "If the other is not too slow in his reaction, he would have to answer: 'Thank you very much, but, excuse me, I now also doubt the correctness of this point of view of yours.'"

Consistent and sincere doubt is lethal to any "–ism"—in particular to relativism and cynicism, which tend to colonize doubt to their own advantage. Let's try to summarize the dimensions of a consistent and sincere doubt:

Doubt is alien to all "–isms" and their true believers, as well as to relativists and cynics. As we saw above, the human condition seems to be determined by doubt about doubt. Truth is not denied or rejected, but it is believed. To echo Musil once more: The voice of truth has a suspicious undertone. In fact, Popper's methodological device of falsification can be extended to a life borne by doubt. True believers found their existence on the alleged rock of an indubitable truth that offers scores of "verifications"—that is, proofs of this indubitable truth. Yet doubters—those who live a life carried by sincere and consistent doubt—search instead for "falsifications"—that is, for dubitable cases and situations. Eventually, in a slow evolutionary process, an individual may come close to a resemblance of truth—or, if you will, verisimilitude (literally, "something resembling truth").

Sincere and consistent doubt is the source of tolerance, as was demonstrated by Castellio in his opposition to Calvin's theocratic terror. Castellio believed in God, but his faith remained linked to doubt, just as his considerable knowledge never lost sight of an inherent ignorance. This worldview is, of course, more than a private and personal attitude. It's a way of life that stands at the cradle of Western democracy. Indeed, doubt is the hallmark of democracy, just as absolute truth (alleged and truly believed) is the hallmark of every type of tyranny. After all, isn't institutionalized opposition, as a component of multiparty government, a countervailing force, and as such the essence of the democratic political system? Governmental debates about the pros and cons of policies are continued in the media and the conference halls of the civil society. If doubt were to come to a final and absolute rest, democracy itself would come to an end—there would remain nothing to be debated! It's in this public space created by systematic political doubt that our civil liberties and constitutional rights are safeguarded. In sum, democracy is unthinkable without sincere and consistent doubt. Conversely, as we'll see in more detail in the next chapter, our existential doubt needs the safeguard and guarantee of a democratic, constitutional state.

CAN DOUBT EXIST WITHOUT FALLING INTO RELATIVISM?

Before we conclude this chapter we want to comment briefly on a "middle position" between relativism and fundamentalism defined in religious terms, then attempt an outline of the prerequisites of any such "middle position" (religious or otherwise).

Whatever criticisms one may make of Max Weber's ideas about the relationship between Protestantism and the genesis of modern capitalism, there's one thing that he understood very well and that few critics have challenged—namely, that Protestantism has had a unique relationship with modernity. This isn't the place to reiterate this argument. Its main features, though, are quite clear: The Reformation, in putting a unique emphasis on the conscience of individuals, laid the foundations of modern subjectivity—and thus of the panoply of rights of the individual as those have been developed and refined since the Enlightenment.

It can't be stressed enough that this historic achievement was unintended—indeed, Luther and Calvin would have been appalled by many features of modernity. And neither reformer can plausibly be interpreted as occupying a "middle position" as defined in this chapter. Enough has been said about Calvin's credentials as a *bona fide* (literally!) fanatic. Luther is a little more difficult to subsume under that category (perhaps mainly because he had a highly developed sense of humor). He may not have presided over the burning of any heretics on the main square of Wittenberg, but his bloodthirsty writings during the Peasant Rebellion and the repulsive anti-Semitism of his later years certainly disqualify him for any humanitarian decorations. Still, it's possible to define a "middle position" on clearly Protestant lines, beginning with an elaboration (again, certainly not intended by Luther) of his pivotal idea of salvation by faith alone (*sola fide*). By definition, faith is not certainty, and thus doubt can be accommodated most easily in a Lutheran version of

Protestantism. Likewise, the Lutheran doctrine of the two kingdoms—that is, earthly and spiritual—makes impossible the kind of theocracy that Calvin set up in Geneva. And Lutheran ethics brought about (directly as well as indirectly) the birth of the modern welfare state in nineteenth-century Germany.

Calvinism, too, had remarkable unintended consequences. One of its fiercest branches (assuredly qualifying for the label of fanaticism) was the Puritanism that was dominant in early New England. But the peculiar American circumstance of a plurality of religious groups inadvertently led to the transformation of the churches into voluntary associations, and thus to religious tolerance and church/state separation.

Probably the most dramatic manifestation of a Protestant "middle position," seeking to bring into balance faith and doubt, was the birth of modern biblical scholarship in Protestant theological faculties in the nineteenth century, especially in Germany. This was a unique case in the history of religion, with professional theologians turning the skeptical discipline of modern historical methods on their own sacred scriptures—with the intention not of attacking faith, but of reconciling it with truths derived from other sources.

Arguably, if one wants to define a "middle position" on religious grounds, it helps to be Protestant. But one doesn't *have* to be Protestant in order to do so: Such an exercise can be undertaken by other Christian traditions (notably by Roman Catholicism in the wake of the Second Vatican Council), by Eastern Orthodoxy (notably in the works of theologians in the diaspora in the West), by Judaism (particularly the core of the highly nondogmatic and skeptical rabbinical method), and by Islam (building on the concept of

Qur'anic interpretation, *ijtihad*). Needless to say, it isn't possible to develop all these points here.

One of Immanuel Kant's books, written in the eighteenth century, has the impressive title *Prolegomena to Any Future Metaphysics That Will Be Able to Present Itself as a Science*. Not that we want to put ourselves in Kant's league, but we would venture to head the final section of this chapter "Prerequisites of Any Future Worldview That Will Be Able to Present Itself as a Middle Position Between Relativism and Fundamentalism." Those prerequisites follow:

1. *A differentiation between the core of the position and more marginal components* (the latter what's been called *adiaphora* by theologians). The practical consequence of this differentiation is to mark the outer limits of possible compromise with other positions. In the modern plural situation, there are strong pressures toward such compromises—in sociology-of-knowledge terms, toward cognitive and/or normative bargaining. For example, Christian theologians may define the resurrection of Christ as *core*, but the other miracles of the New Testament as in principle negotiable. For another example, in the current European debate over the integration of Muslim immigrants into democratic societies, the mutilations and stonings mandated by traditional Islamic law may be deemed nonnegotiable, but the wearing of kerchiefs (*hijab*) in the name of "Islamic modesty" may be negotiable.

2. *An openness to the application of modern historical scholarship to one's own tradition*—that is, a recognition of the

historical context of the tradition. Such a recognition makes fundamentalism difficult to maintain. We've already mentioned the dramatic case of Protestant biblical scholarship, its openness now absorbed by Catholics and some Jews, though as yet very few (if any) Muslims. In the latter case, a *theological*, as against a merely *factual*, differentiation between the portions of the Qur'an that originated, respectively, in Mecca and Medina will be very important for a distinction between core issues and *adiaphora* in Islamic thought (and indeed Islamic practice). This point is obviously more relevant for religious than for secular traditions, though there are secular analogues. The debates within Marxism of the relation between Marx's early writings and *Das Kapital* is a very interesting case.

3. *A rejection of relativism to balance out the rejection of fundamentalism.* Relativism leads inexorably to the cynicism discussed earlier in this chapter. If "anything goes," cognitively as well as morally, the position as such becomes basically irrelevant: If there's no such thing as truth, one's own position becomes a completely arbitrary choice. If relativism is applied cognitively, flat-earth theory has to be given the same epistemological status as modern astronomy—or, for a more timely case, creationism and evolution would have to be given equal stature in a high-school curriculum. Relativism has normative consequences as well: It would argue that the "narrative" of the rapist is no less valid than the "narrative" of his victim.

4. *The acceptance of doubt as having a positive role in the particular community of belief.* We need not repeat what we said about this earlier in this chapter.

5. *A definition of the "others," those who don't share one's worldview, that doesn't categorize them as enemies* (unless, of course, they represent *morally* abhorrent values). In other words, the community of belief must have the ability to live in a civil culture and to engage in peaceful communication with the "others." Manifestly, the absence of such civility leads to disruptive processes in society, ranging from a vituperative climate in public life to violent civil war.

6. *The development and maintenance of institutions of civil society that enable peaceful debate and conflict resolution.* Politically, the liberal democratic state, guaranteeing human and civil rights (notably freedom of religion and freedom of association), is by far the best available political system for enabling peaceful debate and conflict resolution. Even the Jacobin formula, which accepts no intermediary between the individual and the state, is not conducive to the moderation of "middle positions," even if the state is formally democratic. History has shown a need for "mediating structures"—an array of institutions standing between private life and the state. This is what's meant by *liberal* democracy; as political columnist Fareed Zakaria recently reminded us, there are also *illiberal* democracies, which maintain the machinery of competitive elections without their foundation in civil society. What has been happening of late in various countries of the Middle East makes this point clearly.

7. *The acceptance of* choice, *not only as an empirical fact but as a morally desirable one.* This acceptance is not only a matter of allowing individuals to make unconstrained decisions on a wide array of religious, moral, and lifestyle issues (obviously within certain limits—I should be free to follow my religious beliefs but not to practice ritual cannibalism, to choose my "sexual preference" but not if it entails rape). It's also an *institutional* matter—that of accepting a plurality of voluntary associations, again over a wide array of religious, moral, and lifestyle issues.

6

THE LIMITS OF DOUBT

There are cognitive and moral limits to doubt. As we saw before, there are sufficient reasons for *doubt* about doubt. We will address this matter now in greater detail and confront the central paradox of doubt: We can live with considerable uncertainty concerning our cognitive definitions of reality (such as our religious worldview) while at the same time being able to make moral judgments with great certainty. To take but one example (which, unfortunately, has an element of timeliness in the United States at the time of writing): We can say with certainty, regardless of our political or religious orientation, that torture is, at all times and in all places, utterly unacceptable.

AT WHAT POINT, AND HOW MUCH, SHOULD DOUBT ITSELF BE DOUBTED?

One of doubt's primary functions is to defer judgment. Doubt is particularly opposed to hasty judgment, prejudgment, and prejudice. Yet there's a great risk here, and it's precisely at this point that we should realize how problematic doubt can be—or, to phrase it differently, it's at this point that doubt should be doubted. After all, judgments can't be avoided in life (certainly not in politics and in matters of worldview and religion). While a doubter can carefully consider the facts, the possibilities, and the options available before coming to a conclusion and a judgment, eventually choices have to be made and actions taken. The excessive postponement of choice and action can have disastrous consequences, as has been demonstrated time and again in situations of war and serious social unrest. In a sense, not choosing is actually a choice also—and usually a tragic one. To put it concisely, doubt without limits leads to both individual and collective paralysis. The fourteenth-century parable of Buridan's ass is a classic illustration of this.

Sociologically speaking, doubt tends to undermine the certainty of traditional institutions. Institutional certainty is, if it functions well, prereflexive, taken-for-granted, tradition-ally grounded "background" of a society. The type of doubt we're discussing here doesn't *intentionally* destroy this insti-tutional certainty, if only because doubt is averse to radical rebellion and revolution. Nonetheless, at times it questions the taken-for-grantedness of various institutions, subjecting them to closer investigation. Within limits, this sort of doubt serves a useful social function. Indeed, this "debunking" pro-clivity has been the very positive contribution of modern social thought. But doubt with no limits at all—doubt that has abandoned any and all certainty—leads to a fruitless subjectivism in which the individual endlessly reflects about options and ponders all possibilities. As Gehlen argued convincingly, subjectivists of this type usually come to a complete standstill—a state that he called *Handlungsverlust*, or loss of the capacity to act. When doubt undermines the beneficent certitude of an institution, it causes fundamental unrest. This disquiet may admittedly be fruitful in bringing about a fundamental cultural change, and may even be a precondition for artistic or intellectual creativity. However, when anti-institutional doubt deeply penetrates the minds and moods of the population, it can degenerate into chaos and disorder or, as we saw before, end up in cynicism and mindless relativism. Doubt needs sound rationality to keep it under control. Unlike the false certainty of the true believer, doubt is a risky affair.

Cynicism and extreme relativism aren't the only dangers involved here. If applied systematically, doubt can result in despondency, in the loss of hope and action. Doubt, especially doubt about doubt, easily drifts off into desperation. In German, doubt and desperation have the same linguistic stem—*zwei* ("two"). The concepts *Zweifel* (doubt) and *Verzweiflung* (desperation) both indicate that there's a choice between two mutually exclusive possibilities. In medieval German, the word *twi-fla* (which later became *Zweifel*) meant "dual meaning," and that shared root is reflected in the closely related *Verzweiflung*, which can be translated as "complete desperation" or "total despair." Clearly, this sort of doubt is the opposite of cynicism, which is a boisterous celebration of an allegedly omnipresent doubt. Doubt about doubt sliding off into desperation is, of course, no cause of jubilation.

HOW CAN WE ARRIVE AT
MORAL CERTAINTY?

We have now come to the heart of the problem addressed in this chapter: *On what basis can one make moral judgments with great certainty? And how can such certainty be grounded in social institutions?*

The great Rabbi Hillel, an ancient Jewish sage, proposed that the meaning of the Torah could be stated by a person standing on one foot. That meaning, he said, is that one should not do to another what is despicable to oneself (this was actually the first formulation of the Golden Rule). Then Hillel added, "The rest is commentary." One might say that all one's moral certainties can also be summed up while standing on one foot, by one sentence in the constitution of the Federal Republic of Germany: "The dignity of man is inviolate" (*"Die Würde des Menschen ist unantastbar"*). "The rest is commentary." It's instructive to recall the historical context of this sentence. If human dignity hadn't indeed been violated in horrific ways under the Third Reich, the declaration of dignity wouldn't likely have been put into a state constitution. Not always, but very commonly, certainty in moral judgments arises in situations in which one is forced to confront instances of obvious and massive immorality.

Now let's return to the moral judgment mentioned at the start of this chapter: *Torture is, at all times and in all places, utterly unacceptable.* How can one be certain about this judgment?

Generally speaking, there have been four ways in which such certainty has been legitimated—by divine commandment, by natural law, by sociological functionality, and by biological functionality. We find all of these less than persuasive:

1. *Divine commandment.* One can, of course, ground *moral* certainty in *religious* certainty. If one can bring oneself to believe in an absolutely binding divine law, that faith makes moral certainty easy. However, by the same token, it can make such certainty deadly. At our particular juncture in history, this statement hardly needs elaboration. If one does indeed find torture unacceptable, one must then find "proof texts" that proscribe torture in one's sacred scripture and tradition. Unfortunately, this exegetical exercise doesn't always succeed. Furthermore, if one doubts the certainty of religious pronouncements, that avenue is closed.

2. *Natural law.* The venerable tradition of natural law theory maintains that there are moral principles inscribed in human hearts everywhere. In other words, the idea of natural law implies universally. It's a very tempting idea. It's also very difficult to maintain in the face of empirical evidence to the contrary. For example, the condemnation of torture is anything but universal. People have happily tortured each other throughout most of history. A judge in Tudor England, say, or Ming China, would have been surprised to hear that torture is unacceptable: Of *course* you want to torture a suspect— that's how you'll find out what he did. And if you can convict him, you may want to torture him as part of his punishment, to serve as a deterrent—*pour encourager les autres.* In the sense of an empirically available universality, natural law is not a plausible basis for moral certainty, although there may be a modification of the idea that could provide such a basis—natural law as being

progressively revealed in distinct historical stages. We'll come to this possibility shortly.

3. *Sociological functionality.* To base moral certainty in sociological functionality is to say that norms are necessary for social order. That's correct, of course; it's very plausible for certain moral maxims. For example, no society would long survive if it tolerated indiscriminate violence: You take my parking space, and I'll kill you. But that's sociologically intolerable *within* a given collectivity. Such violence may be quite tolerable if it pertains to people *outside* the collectivity: If you (an outsider) presume to park in *our* parking lot, I'll kill you. In the case of torture, opponents would find it difficult to argue that it's intolerable even *within* a given collectivity, and much less so if *outsiders* are targeted.

4. *Biological functionality.* Finally, it has been proposed that norms facilitate the survival of a species in the fierce competition of biological evolution. Let it be stipulated that there's a gene (an instinct, if you will) that makes a mother take care of her newborn child. It's obvious that a species in which this gene is absent has little chance of surviving, while a species that happens to have the gene (presumably as a result of the chaotic game of Russian roulette known as evolution) will have a much greater chance of survival. But torture? Members of a species can torture people within a group without harming their survival chance (though it would have to be done discriminately—say, only to the few individuals who violate a particularly sacred taboo—since if it

were done indiscriminately, the group would commit reciprocal suicide). And the survival chance would clearly be unaffected by torture inflicted on *outsiders*. In any case, biological functionality is no more useful then sociological functionality in providing a basis for moral certainty.

We would suggest a different way of legitimating moral certainty: *Such certitude is based on a historically developing perception of what it means to be human, which, once attained, implies universality.* Put differently: *The meaning of the dignity of humankind comes to be perceived at certain moments of history; however, once perceived, it transcends these moments and is assumed to be intrinsic to human being always and everywhere.* The direction we propose is arguably a variant of natural law theory, but because this book isn't a philosophical treatise, we can't do justice to the complicated question of the extent to which that is or isn't true. We'll return to this point briefly later in this chapter. For now, though, our approach must be elaborated a bit further in the context of the present argument.

What does it mean when we speak of *perception* ("comes to be perceived") in the above restatement? It's a truism to say that, in the course of socialization, morality is internalized in the consciousness of individuals. This phenomenon is conventionally referred to as *conscience*. And the way that conscience is conventionally assumed to "speak" is in the imperative mood: Do this; don't do that. Perhaps this is so in some cases. More commonly, we would suggest, conscience speaks in the indicative mood: Look at this; look at that. In other words, conscience induces specific *perceptions*. These can be both positive and negative. For example, the perception can be of a situation that requires a positive action. This is what the Chinese philosopher Mencius had in mind when he suggested that even the most hardened criminal, if he comes upon a child tottering on the edge of a deep pond, will be moved to pull the child to safety. Alternatively, the perception of some situations induces repulsion. Thus Harriet Beecher Stowe's novel *Uncle Tom's Cabin*, which served to mobilize abolitionist sentiments just before the American Civil War, was not a sermon on the evils of slavery; rather, it depicted scenes from the reality of slavery which were so repulsive that more and more people said, This cannot be allowed to go on.

This latter perception was relative in time and space, of course. Harriet Beecher Stowe was the sister of a prominent Congregational minister (in the best tradition of Puritan morality) in the mid-nineteenth century in the northern United States. Her contemporaries in the antebellum South clearly didn't agree with her, and neither did people throughout most of human history. However, once the perception was firmly in place, it robustly resisted—and continues to resist—relativization: If slavery was wrong in nineteenth-century Dixie, it's wrong in twenty-first-century Sudan. What we observe, then, is a development in the perception of human dignity and of offenses against it from mere opinions ("You and I will agree to disagree about slavery") to universally valid moral judgments ("I condemn your practice of slavery, and I will do whatever I can to stop you").

At various moments in history, there have been voices saying that torture is morally unacceptable. Mencius, for example—representing Eastern thought—may have thought so. In the history of Western civilization, the image of

humanity found in the biblical tradition was a major source
for the developing perception that human dignity precluded
torture. Yet it was only with the Enlightenment that this
perception became generally held in the West and then
was institutionalized in law. Very likely this perception
was induced by the cruelties of law under the *ancien régime*
against which the French Revolution revolted. Voltaire was
an eloquent spokesman for the protest against torture in
France, and he influenced Catherine the Great of Russia.
The Empress Maria Theresa then outlawed judicial torture
in the Habsburg realms. Interestingly, her successor, Joseph
II, proceeded to abolish capital punishment—precisely because it came to be perceived as a particularly heinous case
of torture, however "humanely" administered (the guillotine,
the electric chair, lethal injection—all futile attempts to
camouflage the intrinsic horror). Two books (this time not
novels) were very influential in the abolition of the death
penalty in, respectively, France and Britain—Albert Camus'
Reflections on the Guillotine and Arthur Koestler's *Reflections on
Hanging*. Both men were motivated, not by abstract theoretical considerations, but by revulsion against the horrors
perpetrated by the totalitarian regimes they had experienced
at close quarters: "Look at this," they said, in effect: "It must
not be allowed to continue." No doubt the same perception
motivated the more recent absolute prohibition of the death
penalty within the European Union. (In this, if not necessarily in other matters, the EU is morally superior to the
United States.)

Indeed: "The dignity of man is inviolate. The rest is commentary."

WHAT DOES PHILOSOPHICAL ANTHROPOLOGY BRING TO THE DISCUSSION?

In looking for a plausible legitimization of moral certainty, it's useful to ask what there is about the constitution of *homo sapiens* that makes morality possible in the first place—*not* in the sense of supplying specific moral maxims ("Torture is unacceptable"), but rather in the sense of explaining how human beings can be *moral* beings in the first place. The discipline that asks such questions is known as philosophical anthropology. It can't tell us why torture is unacceptable, but it *can* tell us how human beings are capable of asking and answering moral questions. An analogy might be useful here: Biologists and linguists have been able to show that human beings have an innate capacity for language. This capacity may even have produced "deep structures" of language, setting the parameters within which any specific language (say, Swedish or Swahili) has constructed its lexicon and grammar. We can't directly deduce the specifics of the Swedish language from those "deep structures," however, and we likewise can't deduce Swedish values on torture and capital punishment from philosophical anthropology. Still, it's useful to be aware of the latter if one tries to understand the former.

Philosophical anthropologists try to determine what the components of the human condition are. One fundamental component is the "institutional imperative"—that is, the need human beings have for institutions (traditional patterns of acting, thinking, and feeling, as Emile Durkheim defined them) in order to survive in nature and history. And as Arnold Gehlen pointed out, human beings lack distinct, biologically determined instincts that help them to react adequately to changes in their environment. Institutions are in a sense substitutes for these missing instincts; they help us to react quickly and without much reflection (semi-instinctively, as it were) to changes in the environment. When the traffic light changes from green to red, I brake immediately, without reflection, semi-instinctively. That reaction isn't *truly* instinctive, of course; rather, it's learned behavior. More specifically, it's *institutional* behavior, because motorized traffic is a modern institution with specific values and norms into which we've all been socialized. Such institutional reactions aren't biologically fixed, but historically and sociologically changeable. Moreover, institutions—the family, the church, the school, the university, the workers' union, and so on—can be seen and studied sociologically as if they were "things," as Durkheim emphasized. Yet they're human constructions that change over time—usually slowly, sometimes (in times of revolution) rapidly. There have been and still are theologians and philosophers who believe the institutions to be given by God or the gods. For instance, the theological formula of the marriage ceremony is: "What God has united, let no one separate." But, sociologically understood, marriage is an institutional construction with a particular history, and it can be analyzed as such—and by the same token can be relativized and subjected to doubt.

Another important (indeed, essential) component of philosophical-anthropological analysis is the fact that humans are speaking and communicating beings. Language—first spoken language, then written—is the main instrument for the communication between human beings. Words not only *refer* to things in the outside reality, but *define* them as well, in terms of beauty, utility, danger, succor, and so on—and the opposites of these. That is, spoken words aren't sheer guttural sounds, but carry meanings and values. Connected to the latter are norms—rules of moral behavior—because values demand commitments, which are expressed by norms.

But how, precisely, do these linguistic meanings, values, and norms—the pillars on which social life rests—come about? Again, it won't do to refer simply to metaphysical instances such as "nature" or one or the other "divinity." The answer must be empirical. It lies, we suggest, in the central concept of *reciprocity*.

This concept, which played a definitive role in social psychology, was best explored by the American philosopher George Herbert Mead in his "symbolic interactionism," which he himself preferred to call "social behaviorism." The bulk of our actions in society are *interactions*, Mead noted—actions with fellow human beings, some of whom are "significant others" (parents, children, spouses), others being less "significant" people with whom we interact (casual acquaintances, neighbors, the milkman). According to Mead, the child eventually integrates all these interactions into an abstract notion of society, the "generalized other," which is the repository of internalized norms. The sequence is as follows: "Mummy gets angry if I pee on the floor"—"The milkman gets a bad idea about me if I pee on the floor"—"One does not pee in the floor." (The last formulation of the norm is best expressed by the French word *on* or the German *man*.)

Interactions, Mead argued, often start with an exchange of physical gestures. Someone, for example, swings his fist under my nose. In itself, this is a meaningless gesture that I must try to interpret. It could be a joke, say, or an innocent bantering gesture. However, if the other person's facial expression indicates that the gesture is an aggressive one, I step back and clench my fists. This response is seen by the aggressor as an inimical act on my part, and thus the exchange of gestures develops into a meaningful interaction that has a name—a fight. Now, the crucial dynamic in all this is *reciprocity*. I understood the hostile gesture of the other because I'm able to take on the role or the attitude of the other, as in a kind of prereflexive empathy. The other person—as he interprets my reaction (the clenching of the fists) as a defensive-offensive, hostile gesture—likewise internalizes and interprets my attitude. In other words, through our mutual internalization and interpretation, meaning emerges. An exchange of initially meaningless gestures transforms into a meaningful interaction.

It gets more complicated. In reciprocal interactions, Mead said, we internalize the role or attitude of others, and in doing so we direct our thoughts, emotions, and actions not only toward others "outside" ourselves, but also toward *ourselves* in the internalized role of the other. In other words, we experience an *internalized* reciprocity. Take the interaction between a teacher and her students as an example. The teacher addresses the students sitting before her in the classroom. She talks to *them*, but at the same time she takes the role of a student and addresses *herself* in the internalized role of a student. The same happens within her students, who assume the role of the teacher and address themselves in that internalized role. Reciprocity is thus a mutual taking/internalizing of the role/attitude of the other. It's in this way that meaning (and thus mutual understanding) can emerge in interactions. These interactions are then more than behavioral movements; they're meaningful exchanges that can be given names. In the example above, we speak of "teaching" and "learning" and of understandable, meaningful roles like "teacher" and "student."

Mead added a moral dimension to this theorem of meaningful reciprocity. Reciprocity makes possible an empathy with the others I interact with: I can internalize within my own consciousness the perceptions and feelings of these others—to use a somewhat hackneyed phrase, "I feel their pain." This in turn leads to a feeling of moral obligation on my part: Empathy becomes charged with moral meaning when it leads to the proposition that I should stop inflicting such pain. Clearly, in the case of torture, I condemn it if I put myself, by means of reciprocity, into the victim's shoes. But, equally clearly, the history of torture shows that I can avoid such empathy (be it as a witness to torture or as a torturer myself), if I can manage to deny the victim the status of someone with whom I have a relationship of reciprocity. To stay within the Meadian paradigm, I then deny the victim's status as a potential "significant other"—indeed, sometimes, in extreme cases, deny the status of a human being. The literature on the mind-set of Nazi murderers during the Holocaust provides chilling examples of this phenomenon. Likewise, this is how most slave owners have avoided empathy with their slaves from time immemorial.

IS MORALITY PART OF "HUMAN NATURE"?

These philosophical-anthropological considerations show how morality is possible at all—that is, as a result of the built-in reciprocity without which an individual couldn't be socialized. But the boundaries within which the individual "indulges," as it were, in this reciprocity are socially constructed, and thus they depend on the "significant others" by whom the individual was socialized: A child raised exclusively in a family of professional torturers will have different limits on empathy as compared to a child raised by pacifist Quakers. The respective groups of "significant others" will also determine the nature of the child's "generalized other"— that is, the internalized image of the society to which one has moral obligations (in other words, the people to whom one owes anything in moral terms). In the course of history, the boundaries of reciprocal obligations have shifted repeatedly, sometimes leading to wider inclusion, sometimes narrower. Clearly, then, the capacity for reciprocity is an anthropological constant, but it's subject to all kinds of manipulation. It's definitely useless to argue against torture on philosophical grounds with a torturer who sincerely feels no empathy with and thus no moral obligation to his victims. To convert him, one would have to change his perceptions—*to make him see*. There are cases where this has actually happened. To shift from torture to slavery, a powerful literary depiction of such a conversion is the discovery by Mark Twain's Huckleberry Finn that the runaway slave with whom he's been traveling is a human being like himself and thus should *not* be returned to his master—indeed, should not be a slave at all.

It should be clear by now in what respect our argument

can be placed in the tradition of natural law theory (though, as noted earlier, we won't address the many ways in which contemporary philosophers have struggled with this question, or the various positions in play). To the extent that we're grounding moral obligation (and *ipso facto* moral judgment) in the human capacity for reciprocity and empathy, and given the fact that this capacity is an anthropological constant, we're proposing a natural law argument. Contrary to, for example, existentialist and postmodernist theories, we argue that there's something like "human nature," by definition universal. The most egregious assaults on human dignity, we assert, are possible only by denying or repressing this "nature." And, indeed, the full contours of this "nature" have become visible only over many centuries of human history—not a unilinear progress, though, since perceptions have been lost as well as acquired over time.

Unfortunately, the moral implications of all of us sharing a common "human nature" are all too easily avoided. We've previously mentioned one method of avoidance—denying the humanity of victims. Another frequent method is denying the agency of the victimizer. This has very often taken a religious form: "*I* am not really doing this. I'm only God's *instrument.*" A grisly expression of this avoidance can be seen on many European executioners' swords (such as are exhibited in the Tower of London): "Thou, Lord Jesus, art the judge." In other words, "*I'm* not chopping off your head; Jesus is." The so-called Stockholm syndrome (named after an episode in which kidnapped people began, after a few days of captivity, to identify with the kidnappers, and even defended them) occurs when the victim of beheading shares the delusion: "Thank you, Lord Jesus, for chopping off my head; I fully deserve it." Jean-Paul Sartre called this phenomenon "bad faith," be it on the part of victims or victimizers.

Let it be noted that such "bad faith" can take secular as well as religious forms. Thus the judge pronouncing sentence—not only a sentence of death (where such a barbaric practice still prevails), but *any* sentence—can similarly deny her own agency: "*I* am not doing this. The *Law* is. I'm only an instrument of the Law." For example: "I'm ordering foreclosure of your home. I'm not responsible for your ending up on the street, however, because *I* am not doing this to you. I'm only an instrument of the Law." One can argue that social order, and certainly any system of law, would be impossible without such "limited liability" for those who play certain roles in that social order, such as a judge. However, this doesn't change the fact that limited liability, in this case, involves a big fiction.

IS MORALITY SIMPLY A MATTER OF FOLLOWING PRINCIPLES?

If the preceding argument has merit, it's indeed possible to make moral judgments in an attitude of certainty. That clearly places a limit on doubt. But, as we've tried to show, the certainty applies to only a relatively small number of "clear cases"—pulling back the child about to fall into the pond, stopping torture, helping a runaway slave. Most situations calling for moral judgments are much less clear. Consequently, hesitation and doubt are very much to be recommended. As Oliver Cromwell called out to Parliament, "I entreat you, by the bowels of Christ, consider that you may be wrong!" (The quaint phrase "bowels of Christ," which comes from Paul's Letter to the Philippians, apparently is a synonym for "the mercy of Christ.") Thus there's room for doubt within moral certainty, just as there's room for doubt in fervent faith.

Max Weber's distinction between an "ethic of attitude" (*Gesinnungsethik*) and an "ethic of responsibility" (*Verantwortungsethik*) is relevant to this discussion. The former type of morality has *principled attitude* as its basic criterion. Weber takes Tolstoy and his absolute pacifism as a prototype. He could equally well have taken Gandhi, who was a sort of disciple of Tolstoy's and who claimed nonviolence as an absolute principle, regardless of consequences. There can be a certain grandeur to the principled attitude, but it can also be grossly irresponsible. During World War II, a group of Jews asked Gandhi whether he absolutely opposed resisting Hitler by violent means. Gandhi said yes. They then asked, What if, as a result, they were all killed? Gandhi replied, You could then die in the knowledge of your moral superiority. If the Jews accepted this argument, they were merely stupid. But the fact that Gandhi made the argument showed the irresponsibility of his ideology of nonviolence.

An "ethic of responsibility" is opposed to the above-described ethic: Rather than asking, What is the attitude one should take? it asks, What are the probable consequences of one's actions? If one aims for the right consequences, one then acts out of responsibility even if one gets one's hands dirty. Weber approvingly quoted Machiavelli's statement that a ruler should act for the welfare of his city, even if thereby he imperils the eternal destiny of his soul. (One could argue that Weber here represents a secularized version of Lutheran ethics. But that's another story.)

The debate over abortion, which has raged in America for decades, illustrates the precarious relation of certainty and doubt in the area of moral judgment. Two principles involved in the debate can be affirmed (indeed, *are* affirmed) by both sides with great certainty:

1. That every person has the fundamental right to life (which is why murder is one of the most terrible crimes).

2. That a woman has the fundamental right over her own body (which is why rape is a crime comparable to murder in its violation of human dignity).

The two sides, however, have made very different choices as to which of these two certain judgments applies to the abortion issue. The two labels used in the debate are equally misleading. One side calls itself "pro-life." Yet "life" isn't the issue. Of *course* the fetus is "human life"—so is one's appendix (if one still has it). The real question is whether the fetus is a human person. The other side calls itself "pro-choice." Of *course* a woman has the right to choose what to do with her own body. But the question is where, and when, a woman's own body ends and another person's body begins. After all, a woman doesn't have the right to kill her two-month-old baby (even if she's breast-feeding it). There's a quite different question underlying the debate: *When in the nine-month trajectory of pregnancy does a human person emerge?* The honest answer, in our view, is: *We don't know.* The "pro-life" side, of course, claims to know—at the instant of conception. This view is generally based on theological or philosophical premises that don't convince those not already convinced. Of late the view has also claimed to be based on science, because every fetus at conception already has a specific DNA. But that is likewise unpersuasive: Whatever I am as a person isn't identical with my DNA, and it's *I*, not my DNA, who have certain inalienable rights. The "pro-choice" side, on the other hand, generally avoids the question—it can be very embarrassing, after all—but at least some of its adherents seem to say that the fetus has no rights at any stage of the pregnancy, as in the quite repulsive stance taken by some in the debate over late-term abortion (known by its opponents as "partial birth abortion"). It seems to us that the extreme positions of the two sides are equally implausible—(1) that a person, with all the human rights pertaining to such a one, is present five minutes after conception, and (2) that there is no such person five minutes before birth.

Thus, while in no way denying the aforementioned moral certainties—a person's fundamental right to life and a woman's fundamental right regarding her own body—we seriously doubt that either one or both can be unambiguously applied to the issue of abortion. In this matter, as is the case more often than not, we're faced with the necessity of making a morally sound decision in a state of ignorance about the basic question underlying the issue at hand—in this case, the question of when human life emerges. We must doubt anyone who claims to know the answer in ringing tones of certitude, and we must doubt whatever more or less clumsy solutions we can come up with ourselves. Since we don't know when we're dealing with a yet unborn person (as opposed to a collection of DNA-determined cells), we should proceed very cautiously, favoring a conservative approach. This probably means making abortion solely the woman's prerogative at least during the first trimester, then making abortion increasingly difficult and finally illegal except under extraordinary circumstances. This is actually the legal situation in most European countries, and it strikes us as eminently responsible.

Another interesting case is the current debate in Europe over the integration of Muslim immigrants—a debate that we touched on in our earlier discussion of pluralism and the need for boundaries. It seems to us that there are some issues here about which moral judgments can be made with great assurance, but others about which there's much room for doubt. We can envisage a kind of triage. At one end are issues on which it's possible to be certain—for example, in absolutely proscribing "honor killings," genital mutilation, and the advocacy of violent *jihad*. At the other end are issues on which, it seems to us, one can be fully liberal—for example, in allowing time for prayer to Muslims during working hours, in safeguarding the right to build mosques anywhere (subject to normal zoning considerations), and in allowing the wearing of kerchiefs by women and girls in public places. But between these two poles within the triage, there's a large gray area. Take issues such as the demands by some Muslim parents that their daughters be excused from sporting events with boys, or demands for gender segregation in education generally, or demands for the revival (or introduction) of laws against blasphemy. It seems to us that a cautious, prudent, indeed *doubting* approach to these middle issues is what's called for. In other words, once again, a balance between certainty and doubt.

HOW CAN A CLIMATE OF HEALTHY DOUBT BE MAINTAINED SOCIETALLY?

It's a commonplace of democratic discourse that all beliefs should be protected, except where they advocate or practice assaults on the rights of others. But doubt is a vulnerable and risky thing. It too must be protected against those who would repress it in the name of this or that alleged certainty. We believe that liberal democracy, with a constitution and legal system that protect the freedom to dissent, offers the best system in which doubt can be defended and may even flourish.

Defending doubt isn't everyone's goal, of course. True believers (of any faith and nationality) are prone to attack doubt and doubters, and they try to capture the political arena in order to establish the unbridled tyranny of their professed ideology. Ideologues of tyrannical "–isms" like Fascism, Communism, and Islamism reject civil liberties such as the freedom of speech, and basic legal rights such as the right to a fair trial by an independent judiciary. These safeguards are rejected because the ideologues realize that a democratic, constitutional state institutionalizes procedures that protect doubt. And doubt is what true believers fear most. While tyranny thrives on true ideological belief, democracy is based upon and defends doubt. The multiparty system, which is so characteristic of democracy, guarantees a voice to the opposition, whose task it is to criticize (and thereby throw doubt upon) the policies of the governing party or coalition parties. A particularly honored place is given to doubt in Anglo-Saxon common law: A person is innocent until proven guilty—"beyond reasonable doubt."

Vulnerable doubt needs the protection of a constitutional state, but it's also at the heart of the democratic system. Yet, in view of the permanent threat to democracy on the part of ideological true believers, it's important that the constitutional state and the democratic political system should not be subjected to doubt. This is a remarkable paradox: In order for doubt to exist, it needs to shield the constitutional state and the democratic system from doubt.

This leads to still another paradox: Safeguarding democratic institutions from doubt can easily lead to an absolutizing of democracy and the constitutional state. Indeed, there's the distinct possibility (and occasional reality) of *democratism*, while the absolutizing of the constitutional state can cause the emergence of *constitutionalism*. In democratism, one wants to impose the plural party system and free elections on societies that in many respects are still premodern, radically traditional, and partitioned in terms of ethnicity and/ or religion. One can legitimately wonder if an economically and technologically underdeveloped country is ripe for the democratic system. Democratism can easily be counterproductive and exacerbate situations of underdevelopment, corruption, poverty, and misery. Many postcolonial countries in Africa demonstrate political, economic, and sociocultural disasters brought about by democratism. As to constitutionalism, Germany is experiencing the concept of *Verfassungschauvinismus*, or "constitutional chauvinism," as a kind of surrogate for traditional (and, in that country, historically suspect) nationalism. Constitutionalism argues for the almost sacred and therefore indubitable character of a nation's constitution. There's nothing wrong with a reliance on the norms and rules of constitutional law, of course, but if this reliance evolves into an inflexible and uncritical belief in the allegedly sacred elements of the constitution, it becomes an ideological constitutionalism that's fiercely inimical to doubt. In the United States, for example, doubt about the legitimacy of the death penalty is often silenced by a simple reference to the nation's constitution. The Netherlands shows the very opposite of constitutionalism: In that country, laws and judicial verdicts *cannot* be constitutionally tested.

Let's recapitulate. We have investigated the moral limits of doubt. We have exempted the violation of basic human rights from doubt. We have undergirded our conclusions by the basic anthropological mechanism of moral reciprocity, as outlined in Mead's "symbolic interactionism"—emphasizing, however, that this basic anthropological given doesn't lead to an empirically available, universal moral consensus and that the full implications of this reciprocity develop gradually over long periods of history.

Doubt, as we have seen, is a matter of high risk. It needs political and sociocultural fortification. Without falling into democratism or constitutionalism, the liberal democratic state, with its panoply of constitutional and legal protections of liberties and rights, is, in our view, the most plausible guarantor of cognitive and moral doubt, at least under modern conditions. (The benevolent despotism of the eighteenth century is hard to reconstitute today. Despots are rarely benevolent; however, benevolence is more likely to be institutionalized under democratic conditions.)

It's not our intention here to deny anyone the right to doubt the institutional arrangements of democracy and to freely express this doubt, as long as there's no active attempt to overthrow the democracy. But those of us who cherish democracy will seek to quiet such doubt within ourselves when true believers, of whatever ideological coloration, threaten the very existence of the democratic order. After all, the highest praise of doubt is when it itself is subject to doubt while the conditions that protect it are under attack.

7

THE POLITICS OF
MODERATION

Compared to persons wrestling with doubt (often multiple doubts), true believers have a considerable advantage. Doubters tend to hesitate, to deliberate. True believers, on the other hand, don't have to do anything but act. They have the self-confidence of absolute conviction, and—while they may have to think about this or that tactical direction—they *know* which strategy is the right one because it's determined by their absolute conviction. Put differently: True believers not only work devotedly for whatever their cause is, *they have nothing else to do.* Doubters typically have many other things to occupy them—family, job, hobbies, vices. This is what Oscar Wilde had in mind when he said that the trouble with socialism is that it takes away all your free evenings.

WHAT IS MEANT BY "POLITICS
OF MODERATION"?

In the preceding chapters, we have discussed the religious and moral implications of doubt—that is, of a middle position between the equally undesirable extremes of relativism and fundamentalism. There's also an important political implication: Our position calls for *a politics of moderation*. To practice such politics, one must overcome the aforementioned advantage of fanaticism. One must cultivate all those interests that cannot and should not be politicized. In 1940 Simone Weil wrote a letter to a young girl about the importance of doing schoolwork in a time of war. Weil, admirable though she was, didn't have a great sense of humor. But humor is useful in such circumstances. Fanatics rarely have a sense of humor. Indeed, they see humor as a threat to their alleged certainties. Humor generally debunks certainties and at the same time fortifies those who oppose fanatics. This is why jokes flourish under conditions of political oppression. The Soviet Union and its satellites were a breeding ground for a multitude of jokes, both debunking the various Communist regimes and encouraging those who opposed them.

Those who resist fanaticism need to do so without becoming fanatics themselves. This doesn't mean, however, that they should be anything less than resolute in their political actions. Three remarkable human figures stand out from the struggle in South Africa against apartheid and against the fanatical ideology that sustained it—Nelson Mandela, Helen Suzman, and Desmond Tutu. Each of these oppositionists practiced a politics that was decidedly resolute, yet each of them was moderate in personal views and demeanor.

Helen Suzman, who at the time of writing was celebrating her ninetieth birthday, is perhaps the least well known of the three in the United States. For many years she was the only anti-apartheid member of the South African Parliament—this during a period of savage repression. Despite this, she had a keen sense of humor, which at times she turned against her political opponents. On one occasion she addressed the cabinet and suggested that they might benefit from visiting a black township to see the conditions under which people lived there, but then proposed that they should do so disguised as human beings. On another occasion she spoke to the official opposition party, whose attitude toward apartheid policies had been very limp. She said that often she'd looked at the opposition benches in Parliament, hoping in vain for a glimpse of a shiver trying to climb up a spine.

A joke, more succinctly than an analytic treatise, can debunk an entire ideology. Take the following example from the plenitude of Soviet-era jokes:

What is it when there's food in the cities but no food in the countryside?
The Trotskyite leftist deviation.
What is it when there's food in the countryside but none in the cities?
The Bukharinite rightist deviation.
What is it when there's no food in the cities and no food in the countryside?
The correct party line.
And what is it when there's food in the cities *and* food in the countryside?
The horrors of capitalism!

Doubt need not lead to paralysis. Moderation need not morph into yet another version of fundamentalism. The politics of moderation depends on a balance between a core certainty and many possibilities of action, none of which has the quality of certainty. And that core certainty has to do with the freedom and the rights of the individual—a certainty which, as we proposed in the preceding chapter, can be proclaimed while one is standing on one foot.

WHAT IS THE FREEDOM TO WHICH ALL HUMAN BEINGS ARE ENTITLED?

The freedom we envisage here is not the "negative freedom" (as British philosopher Isaiah Berlin defined it) of classical European liberalism—that is, not the freedom *from* oppressive state control, but rather the "positive freedom" *to act* creatively in all spheres of life. Of course, both types of freedom can be radicalized and thus perverted. Negative freedom can come to mean freedom from *all* restraints whatsoever; in that sense, in every convinced advocate of liberalism slumbers an anarchist. And positive freedom can come to mean an untrammeled individualism that recognizes no standards beyond its own preferences—which is yet another version of anarchism.

In history, the discovery of freedom and the discovery of the dignity of every human person have gone hand in hand. Both discoveries can be found in different cultures. Certainly, the three great monotheistic traditions—Judaism, Christianity, and Islam—contain the image of humanity as God's creation and as ultimately responsible to God. This implies the notion of equality—*all* human beings exist in confrontation with God—and *ipso facto* the recognition of the freedom and dignity of every person. In Christianity these perceptions, as derived from the Hebrew Bible, merged with ideas of individual dignity derived from Greek philosophy and Roman law. Needless to say, it took many centuries before the full moral and indeed political consequences of these beliefs were realized.

Similar (if not, in some cases, identical) ideas can be found outside the context of the above-named "Abrahamic" traditions. The Hindu idea of the eternal self, which in the Upanishads is seen to be identical with the divine ground of all reality, certainly implies the dignity of every person over and beyond the hierarchies of any particular incarnation. It's this image of the self which is intended implicitly when Hindus greet each other with a bow and folded hands. The Buddhist injunction to be compassionate to all "sentient beings" has a similar implication (although classical Buddhism denies the reality of the self). Likewise, Confucian thought ascribes individuation at least to individuals who have successfully cultivated themselves. And traditional African thought is grounded on the value of *ubuntu*, of treating all human beings with kindness. Thus it would be an ethnocentric mistake to propose that only in Western civilization can one find the ideas of the freedom and the rights of all human persons (in the classical American phrase, "irrespective of race, color, or creed"—a phrase recently augmented in terms of "gender or sexual orientation"). All the same, one can make a simple empirical statement: *Only in Western civilization has this perception of what it means to be human been institutionalized in polity and law.*

We're not advocating the classical Enlightenment idea

of progress. There's no *overall* progress to be demonstrated empirically in history. Moral insights develop over time, to be sure—but they can also be lost again. While there's no overall progress, there are specific *progresses*. We would strongly maintain that the institutionalization of the freedom and dignity of every individual is such a progress. If sociology has anything to teach us, it is what Anton Zijderveld has elsewhere called the "institutional imperative." Beliefs, ideas, and values may pop up in various places and at various times. But they will be merely transient phenomena unless they're embodied in institutions. Only then can they be internalized in consciousness and consequently be transmitted from one generation to another.

At this point, we must insist on a proposition that today is decidedly heterodox in "progressive" circles: When it comes to the institutionalization of freedom, Europe occupies a unique position. It is from that location that these institutions have spread to other parts of the world and now are paid lip service almost everywhere (including in countries where in fact very different institutions prevail). As we make this point, however, we must emphasize two additional points:

1. To ascribe this preeminence to Europe doesn't in the least imply that, in its actual history, Europe has been a moral exemplar. Of course not. Anyone maintaining the latter view can be countered by a single word: Auschwitz. Any civilization that created that unspeakable horror cannot claim moral superiority. But this humbling insight doesn't change the fact that it was in Europe that the ideas of freedom and human dignity

were first translated into institutions, even though the spirit of these institutions was frequently violated, often in terrible ways.

2. The recognition that similar ideas can be found in non-European cultures doesn't negate the fact that other cultures failed to produce the institutions that translate these ideas into empirical realities of everyday life, not just for an elite (however defined) but in principle *for everyone.* Islam undoubtedly contains the notion of the equality of all human beings before God, and that belief has in fact overshadowed traditional hierarchies in many places; nonetheless, Islam has coexisted almost everywhere with institutions that negate equality. While the most sublime versions of Hinduism and Buddhism may be said to imply an underlying dignity inherent in all human beings, in practice that dignity has been ascribed only to those who have succeeded on some path of spiritual perfection. And, just as with Islam, in Hinduism and Buddhism the sublime ideas have coexisted with institutions that ongoingly negate these ideas. Mention of Hindu caste should suffice here. Confucianism does indeed have a sublime conception of the individual—but only the individual who has successfully cultivated himself (classical Confucianism, regrettably, did not envisage that women could attain the sublime cultivation).

Nevertheless, the presence in non-European cultures of ideas of individual freedom and dignity helps to explain why the European institutions realizing these ideas have gained such ready assent almost everywhere they've penetrated.

CAN HUMAN FREEDOM AND DIGNITY
BE INSTITUTIONALIZED?

Institutions formalizing human dignity and freedom exist
in what could be described as the democratic triangle. The
three points of the triangle are the state, the market economy, and civil society. The desirable balance among these
three institutional complexes continues to be a matter of
practical and ideological disagreements. Indeed, these disagreements define the boundaries of "right" and "left" politics. But an erosion of any point in the triangle undermines
the other two points as well—if not immediately, then in the
longer run.

Let's look at the three points of the triangle in turn. The
state, of course, is at the core of the definition of democracy.
But we must emphasize that what's at issue here is *liberal*
democracy—that is, a political system in which the proce-
dural mechanics of democracy (uncoerced elections, governments changing as a result of elections, the right of citizens
to organize in order to compete electorally) are combined
with firm guarantees of individual rights and liberties. It's
good to be reminded (as the U.S. reading public was recently by journalist Fareed Zakaria) that there are also *illiberal*
democracies—governments where the procedures of freedom are in place, but without the aforementioned guarantees. In many countries there are popular majorities in favor
of barbaric practices, and if the democracy in such a country
offers nothing more than procedures through which the
majority takes political power, the favored barbarities will be
enacted by democratically impeccable means. Capital punishment, which is still practiced in a number of democratic
states (the United States among them), is an example of a

democratically endorsed barbarity. And yet capital punishment, along with many other barbarities, has been abolished
by *non*democratic governments in quite a few countries—
even, for a while, in Czarist Russia. In other words, there can
be liberal despotisms. Such liberality, however, depends on a
political elite with liberal views—a type of elite that, at least
under modern conditions, is a scarce commodity.

Quite apart from any democratic philosophy, there's a
very high correlation between democracy and the protection of elementary decencies in a society. This generalization can be put in elegant terms of political theory: Power
is corrupting, which is why bastards tend to head governments. Democracy doesn't change that, but it ensures that
the bastards can be thrown out periodically and that there
are limits to what they can do while they're in power. This
isn't exactly a ringing endorsement of democracy, but it will
do for a sober preference in favor of democracy.

The preference for democracy relates directly to the topic
of doubt, which is at the core of our argument. Essential to
a parliamentary democracy is the role of the opposition.
The essence of that role is, quite simply, to throw doubt
on the legislative and policy initiatives of the government.
This is why a democratic system must carefully guard the
legitimacy and the rights of the opposition, as is eloquently
expressed by a phrase used in the British parliament—"Her
Majesty's Loyal Opposition"! In the absence of this kind
of safeguarding, a parliament is likely to be reduced to the
status of an applause machine.

These considerations should discourage people from
embracing democratism as an ideology that aims to in-
stall democratic regimes everywhere in the world, even in

countries where there's little likelihood of the citizenry's pursuing liberal policies. Recent adventures in American foreign policy offer a distressing education in this proposition. Thus democracy *per se* is an implausible candidate for passionate commitment. But the liberal values that democracy should ideally promote are a different matter—values of freedom, human dignity, human rights. Even as one practices the politics of moderation—which we've defined as politics that can accommodate doubt—one can be passionately, indeed immoderately, committed to these values. As previously noted, this commitment especially comes to the fore when freedom and its moral components are threatened.

Now for the second point in the democratic triangle. If democracy presupposes the free political actor—the citizen—the market economy presupposes the free economic actor. The key word in both cases, "free," indicates that both systems are based on the notion of individual rights. The relationship between the first two points of the democratic triangle has been the subject of intense debate for a very long time. In contemporary parlance (the two terms have had different meanings in other times), the "left" tends toward the state in the triangle, and the "right" toward the market. Both tendencies can be carried to extremes that unbalance the triangle—the "left" to an oppressive statism, the "right" to the anarchy of unregulated competition. The *reasonable* versions of "left" and "right" eschew these extremes.

There's an ideology of the "right" (again in contemporary, especially American, terms) that sees democracy and the market economy as reverse sides of the same coin—that is, as coextensive and mutually dependent social arrangements. This isn't quite correct empirically. There can be democracies that effectively destroy the market ("democratic socialism," if you will). There can also be market economies directed by nondemocratic regimes. An empirically more nuanced view is indicated: The market economy is a necessary but not sufficient condition of democracy. What's more, the market economy, once introduced, has a democratizing effect over time, though that effect is neither inexorable nor irreversible. It seems to us that a politics of moderation will have such a nuanced view of the matter. (Obviously, this isn't the place to develop this view in detail.)

Finally, the third point of the democratic triangle is civil society—that is, the variety of institutions that mediate between the lives of individuals and the megastructures of a modern society, including the state and the economy. Religious institutions are a very important example of such mediating structures, which is why one can plausibly argue that freedom of religion is a fundamental right, not only for the sake of religion but for the health of a democratic order. The institutions of civil society set limits to the power of both state and market, and they're ultimately necessary for the survival of both. Conversely, civil society survives best under conditions of a democratic state and a market economy. (Again, we can't develop this view any further here.)

Broadly speaking, the three points of the democratic triangle—the aforementioned institutional aggregates of the state, the market economy, and civil society—are related to three modern ideologies—liberalism, socialism (in its democratic version), and conservatism (also in its nonauthoritarian version). We would argue that a politics of moderation is possible within each of these ideologies. Moderate liberals and moderate socialists (social democrats, if you will) understand the limits of both state and market, refusing to absolutize either. Conservatives tend toward glorifying civil society (Edmund Burke's "little platoons," many of them animated by traditional values), but moderate conservatives understand that, under modern conditions, civil society thrives most exuberantly in a democratic state and a market economy. Conversely, each of these three ideologies can be radicalized—liberalism toward an absolutist understanding of the market (characteristic of so-called libertarianism), socialism toward a totalitarian control of all the institutions of society, and conservatism toward a reactionary (and futile) project of going back to this or that version of traditional society. To achieve the politics of moderation that we're proposing here, people must resist the radicalizing trends within every one of the above three democratic ideologies.

HOW DOES AN ETHIC OF MODERATION WORK?

What we've tried to do here is to show that there can be a middle position between relativism and fundamentalism, not only in religion and morality, but also in politics. The politics of moderation is precisely such a middle position. And, despite many doubts among its adherents, this position can be held with passionate commitment. As we've seen, all radical ideologies tend to produce fundamentalists—true believers, who want to establish taken-for-granted dominance of their worldview, preferably in the society as a whole, or at least within a sector of society under their control. Likewise, there are political relativists, who want to divorce politics from any and all moral truths, because such truths are unavailable or even undesirable. An example of this are radical "multiculturalists," who maintain that all cultures are morally equal and who, therefore, will tolerate any barbarity (at home or abroad) because supposedly it's part of this or that culture.

The politics of moderation needs its own ethic. Think back, for a moment, to Max Weber's two types of ethics: an ethic of attitude and an ethic of responsibility. Fundamentalists always gravitate toward the former (sometimes rendered into English as an "ethic of absolute ends"): Doubt is excluded, because all basic questions have already been answered within the taken-for-granted worldview. Political moderates, on the other hand, gravitate toward an ethic of responsibility. They understand that there are few certainties in the realm of politics, except for the one that can be stated while standing on one leg. Therefore, there are no absolute guides to action, which means that the probable consequences of action must be assessed, carefully and pragmatically, as best one can. After that assessment, the political moderate can proceed to the necessary action, shelving doubts and hesitations "for the duration." Relativists, for their part, have a sort of ethic too. It's the ethic of "anything goes," brilliantly exemplified by so-called postmodernism. Eventually this ends up as an ethic of nihilism. Paradoxically, doubt is repressed at this extreme as well—namely, doubt as to whether there may not be some binding truths after all. It turns out that even postmodernist theorists can be quite fanatical. The fanaticism naturally comes out if anyone dares to question postmodernists' theories.

To concretize the sort of political stance that we're proposing here, a stance based on an ethic of moderation, we will now return briefly to two issues currently much in the public eye, respectively in America and in Europe—capital punishment and the integration of immigrants.

Europeans are (we think, rightly) shocked by the stubborn survival of the death penalty in America. There are a number of reasons for its persistence—the cultural heritage of the frontier, the much stronger influence of religion in the United States as compared with Europe (a factor because, regrettably, people who believe in life after death are more inclined, in Luther's words, to "practice Christian neighborly love" by helping some neighbors to move from this world to the next), and the more democratic character of the United States (a factor because the Western secular elite tends to be opposed to the death penalty—and that elite is more influential in Europe).

Be that as it may, there has been a steady decline in the number of Americans favoring the death penalty, from a strong majority a short time ago to just under one-half now. It isn't clear how this shift is to be explained. Recently, though, the issue has been propelled into public awareness by two developments. The first is the discovery, by means of DNA analysis, that a number of individuals had been sentenced to death for crimes they didn't commit. These individuals were subsequently released from death row. While there's been no conclusive evidence that any innocent person has been executed, the cases of DNA-based acquittals of death-row survivors make it likely that there *have* been wrongful executions. The other development is a number of challenges to the prevailing method of execution—death by lethal injection. Although proponents of lethal injection maintain that this method is more humane than other methods, such as hanging or electrocution, challenges by lawyers acting on behalf of death-row inmates have presented evidence that, contrary to the official view, execution by lethal injection as presently practiced is actually very painful indeed. If that's so, the legal argument goes, lethal injection is a violation of the constitutional prohibition against "cruel and unusual punishment." The U.S. Supreme Court has agreed to take up one such challenge, from a case in Kentucky. Pending a decision, at the time of writing, there has been a *de facto* moratorium on executions in the United States.

We have insisted before that capital punishment is to be rejected because it is, in and of itself, a gross violation of human dignity. Obviously, it's less barbaric if executions are carried out by lethal injection than by the prolonged inflic- tion of torture or by burning someone alive. Nevertheless, *there is no humane way of carrying out a death sentence.* And yet the Supreme Court, even if it decides in favor of the plaintiff, will do so on narrow grounds; it may decide, for example, that *this particular method of execution* (the so-called cocktail of drugs) is constitutionally unacceptable. It's highly unlikely that the Supreme Court would rule against capital punish- ment as such. There will then be no bar prohibiting states that retain the death penalty (still a majority of them) from finding a different combination of drugs, which will then be deemed to be more painless and therefore not in violation of the constitutional proscription.

How is one to view these developments? If one believes in the unacceptability of capital punishment, as we do, the developments are in principle wrong. We wouldn't mind putting this principle in the sort of absolute terms typically associated with an ethic of attitude. Of course it's worse if an innocent person is executed. Of course it's worse if the execution involves the infliction of great pain. But capital punishment is unacceptable even if imposed on a person guilty of the crime in question. And it's unacceptable even by a method deemed to be painless. Capital punishment is intrinsically, unalterably an assault on human dignity, and it would be that even if no innocent person were ever executed and even if a truly painless method of execution could be designed. An ethic of attitude would then suggest that the recent developments should be rejected in principle, and that the death penalty should be solemnly declared to be in opposition to the basic values of American democracy. The strategy of those opposed to the death penalty would then be to insist on immediate and total abolition, both by legislation and in the courts—no matter that such a strategy would very likely fail in the United States, given the present climate of opinion among legislators and judges.

Here's where an ethic of responsibility would suggest a less absolutist strategy, even if such a strategy is quite distasteful morally. The immediate goal must be to reduce executions as much as possible. A moratorium is better than nothing. And as penal authorities continue their futile search for a painless method of execution, public opinion might begin to grasp the fact that there's *no* humane way of executing people, innocent or not. Who knows—perhaps this insight might even penetrate into the minds of the nine members of the Supreme Court who are supposed to represent the acme of judicial wisdom in America.

Let's shift focus for a moment. As we've noted earlier, an urgent issue being discussed in every country of the European Union is the integration of immigrants, especially those

coming from countries remote from the European experience. There's been a definite movement in recent years away
from the "multiculturalist" ideology that was in vogue in
the last decades of the twentieth century. Multiculturalists
argued that immigrants have every right to maintain their
original culture and that, except for refraining from illegal
activity, they should not be expected to integrate with the
culture of the host country—not even if and when they
become citizens. Not surprisingly, this movement away
from "multiculturalism" can take rather ugly forms—racism,
xenophobia, even anti-immigrant violence. The hold of multiculturalism over elite opinion creates difficulties for anyone
who wishes to dissent from this view. Dissenters are typically perceived as racist, ethnocentric, and the like.

In the arena of immigrant culture, as elsewhere, there's a
plausible moderate position between the extremism of tolerating everything and tolerating nothing. It should be clear
that, in a democracy, people should have the right to pre-
serve their family heritage—language, religion, mores. But it
should also be clear that a society has the right to maintain
its indigenous, historical culture and that newcomers, if they
wish to become part of that society, should affirm a definite
degree of loyalty to the indigenous culture. The obvious
problem is where to draw the line between tolerable and
intolerable difference.

In recent years, especially since the violent intrusion of
radical Islam into the West, cultural concern has mainly been
focused on Muslim immigrants. There continue to be Europeans who insist that immigrants should be forced to make
no accommodation to the host culture except the minimal
one of obeying the law. Others insist on total enculturation
of immigrants. Again, though, there's a middle ground. As

we've already noted, we imagine a sort of triage—items of the original culture that are clearly unacceptable (for example, "honor killings" of women), items that are clearly acceptable (such as respecting the religious obligations of Muslim employees), and items that are in a gray area.

It's the gray area that's problematic, of course. Should the authorities interfere when mosques, while not directly fostering violence against infidels, teach doctrines that define all infidels as enemies of the "true religion"? (Admittedly, there's a thin line between words and deeds.) While Muslims have the right to have their places of worship in public spaces, is there legitimate ground to object if a mosque is planned directly across from a Christian cathedral that has for many centuries defined the physical character of a particular city? The list could be easily expanded, as our examples in earlier chapters indicate. There are no clear-cut answers to these questions. They should be approached in a spirit of openness, pragmatism, and respect for the rights of both immigrants and members of the historically indigenous group.

The underlying question here is how a community defines its boundaries of belonging. If there are no boundaries at all, there will be no community at all. Every "we" implies a "they," but the moral and political question is how "they" are defined and then whether they're treated with respect for their human dignity.

The boundaries of belonging: Who are "we" and who are "they"? Many people unthinkingly define this dichotomy in terms of international sporting events. Thus many Britons were shocked when, at a recent cricket match between Britain and Pakistan, British-born spectators of Pakistani descent cheered for the Pakistan team. In a somewhat similar case, there was widespread resentment in the United States when

Mexico extended the right to vote to people of Mexican nationality north of the border, regardless of whether these people were legal or illegal residents in the U.S. To make matters worse, the Mexican president said in a speech that Mexico didn't stop at its international borders.

The boundaries of belonging: One of the most inspiring episodes from Nazi-occupied Europe was the rescue of the entire Jewish population of Denmark in a concerted and very well organized action by every sector of Danish society, from the king on down, by which several thousand Jews were taken across to neutral Sweden under the noses of the Gestapo. After the war, a delegation of American Jews visited the Danish prime minister. The leader of the delegation said, "We come to thank you for what you did for our people." The prime minister replied, "We didn't do anything for *your* people. We did it for *our* people." What he meant, of course, was that the rescue operation hadn't been undertaken in order to do something for "them"—that is, for outsiders; on the contrary, the rescued Jews weren't "they," but rather were "we"—full members of the Danish national community. A contrary (admittedly much less dramatic) example of how boundaries are drawn comes from present-day Germany: With the best of intentions, liberal groups in that country have undertaken campaigns to oppose "hostility toward foreigners"—including among the intended beneficiaries "foreigners" who were born in Germany and are German citizens. The same ambiguity can be found in programs to encourage "dialogue between Germans and Jews"—implying (no doubt unintentionally) that these are two contradictory identities.

* * *

If the argument of this book has merit, the values of liberal democracy, while relative in terms of particular historical developments, nevertheless can legitimately claim universal authority. We cannot assert liberties and rights for ourselves while denigrating this assertion to the level of mere opinion or preference—as if one said, "I condemn torture, but I respect your right to disagree," in the same tone as one might say, "I prefer Mozart, but you're free to prefer Beethoven." The politics of moderation steers clear of both relativism and fundamentalism. Yet it can be inspired by real passion in defense of the core values that come from the perception of the human condition that we have tried to describe in this book. Our praise of doubt in no way detracts from such passion.

INDEX